THE CACTUS WREN

Cover photo: A Cactus Wren and its nestlings.
Harry L. and Ruth Crockett, Phoenix, Arizona.

THE
CACTUS
WREN

Anders H. Anderson
and
Anne Anderson

THE UNIVERSITY OF ARIZONA PRESS
Tucson, Arizona

About the Authors . . .

ANDERS H. AND ANNE ANDERSON developed an amateur interest
in ornithology which led them to a 30-year study of the Cactus
Wren, a familiar bird in Tucson, Arizona, and its environs. Anders
Anderson was born in Sweden, and came to the United States
with his parents at the age of four. His wife Anne was a Minneapolis native — he an electrician by profession, she a teacher.
In addition to this book, they have made numerous other contri-
butions to ornithology, and are well-known to Arizona bird stu-
dents as experts in the field.

5-23-73

THE UNIVERSITY OF ARIZONA PRESS

I. S. B. N.-0-8165-0399-0 cloth
I. S. B. N.-0-8165-0314-1 paper
L. C. No. 72-77133

1765160

TO

HENRY AND FANNY

(HM-1 AND HF-2)

Table of Contents

List of illustrations and tables viii

Foreword xi

Preface xiii

Introduction 1

1. Problems and methods 3

2. The habitat 7

3. Winter behavior and roosting nests 18

4. Vocalizations 31

5. Pair formation and establishment of territory 34

6. The territories from year to year 51

7. The breeding nest 60

8. The start of laying 73

9. The eggs of the Cactus Wren 83

10. Incubation 91

11. The role of the male: secondary nests 97

12. The nestlings 102

13. Development of nestlings 111

14. From fledging to independence 125

15. Nesting success 141

16. The physical environment and Cactus Wren survival . . . 148

17. Replacement and dispersal 159

18. Interspecific relationships: the Curve-billed Thrasher . . . 168

19. Interspecific relationships: other birds, and predation . . . 188

20. Discussion 196

Summary 201

Appendix I: Distribution of the Cactus Wren,
Campylorhynchus brunneicapillus 211

Appendix II: Chronology of the nomenclature of the
Cactus Wren, *Campylorhynchus brunneicapillus* 212

Literature Cited 217

Index 221

ILLUSTRATIONS

1.1 Locations of our three study areas in the vicinity of Tucson, Arizona. ... 2

2.1 The Kleindale Road, Tucson, Arizona, study area in 1941. 8

2.2 The Kleindale Road, Tucson, Arizona, study area, looking north
     from lot 7 in 1938. 10

2.3 The Kleindale Road, Tucson, Arizona, study area, looking south
     through lot 7 in 1956. 11

2.4 Santa Rita Experimental Range, 35 miles (56 km) south of Tucson,
     Arizona, 19 December 1954: cholla association. 12

2.5 Santa Rita Experimental Range, 35 miles (56 km) south of Tucson,
     Arizona, 19 December 1954: wash association. 13

2.6 Saguaro National Monument, east of Tucson, Arizona. Looking
     north into quadrat E2, 11 December 1966. 15

2.7 Saguaro National Monument east of Tucson, Arizona. The saguaro
     grove in quadrat C4, 11 June 1968. 15

2.8 Saguaro National Monument east of Tucson, Arizona. Looking north
     into C3 cholla grove, 11 December 1966. 16

3.1 Locations of roosting nests of a pair of Cactus Wrens in lots 6
     and 7 on Kleindale Road, Tucson, Arizona. 21

3.2 Common variations in the shape of Cactus Wrens' nests. 23

3.3 Location of numbered cholla cacti in lots 6 and 7 along Kleindale
     Road, Tucson, Arizona. 24

3.4 Movements of Cactus Wren in search of roosting nest site on 7
     December 1941, in lot 7 on Kleindale Road, Tucson, Arizona. 25

5.1 Singing stations of HM-23 in January and February of 1944 in the
     Kleindale Road area, Tucson, Arizona. 42

5.2 Paths travelled by Cactus Wrens during a boundary dispute on
     13 July 1947, on Kleindale Road, Tucson, Arizona. 45

6.1 Territories of Cactus Wrens in 1942, 1945, and 1947, in the Kleindale
     Road, Tucson, Arizona, area. 50

6.2 Territories and first brood nests in 1953 in pasture 5 of Santa Rita
     Experimental Range. 53

6.3 Saguaro National Monument, Arizona. Number of territories of the
     Cactus Wren and Curve-billed Thrasher from 1963 to 1968. 55

6.4 1963 Cactus Wren breeding nests, Saguaro National Monument. 56

6.5 1964 Cactus Wren breeding nests, Saguaro National Monument. 57

6.6 1965 Cactus Wren breeding nests, Saguaro National Monument. 57

6.7 1966 Cactus Wren breeding nests, Saguaro National Monument. 58

6.8 1967 Cactus Wren breeding nests, Saguaro National Monument. 59

6.9 1968 Cactus Wren breeding nests, Saguaro National Monument. 59

7.1 Cactus Wren nest in jumping cholla, Saguaro National Monument. 64

7.2 Cactus Wren nest in palo verde, Saguaro National Monument. 64

7.3 Cactus Wren nest in crotch of saguaro, Saguaro National Monument. .. 65

7.4 Saguaro stump, Saguaro National Monument, Arizona, used for
     Cactus Wren nest location. 66

10.1 Incubation attentiveness at nest 19C in lot 7 on Kleindale Road,
     Tucson, Arizona. 94

11.1 Nest building sequence of Cactus Wren on Kleindale Road,
     Tucson, Arizona. 99

11.2   Cholla 17 on 1 March 1959, Kleindale Road, Tucson, Arizona. 100
13.1   Nestling of Cactus Wren, Kleindale Road, Tucson, Arizona. 113
13.2   Mean daily weights of Cactus Wren nestlings on Kleindale Road,
         Tucson, Arizona. 122
13.3   Mean daily length of wing and tail of Cactus Wren nestlings on
         Kleindale Road, Tucson, Arizona. 123
16.1   Length of time that adult Cactus Wrens, banded from 1939 to 1962,
         remained in the Kleindale Road, Tucson, Arizona, study area. 153
16.2   Survival in days after fledging of banded Cactus Wren nestlings,
         Kleindale Road, Tucson, Arizona, study area. 155
16.3   Monthly count of Cactus Wren nests, Saguaro National Monument,
         east of Tucson, Arizona. 157
17.1   Sequence of replacements of mates in Cactus Wrens in the Kleindale
         Road, Tucson, Arizona, study area. 160
17.2   Cactus Wren, Saguaro National Monument. Nesting sequence of
         CM-54 and CF-66. 163
18.1   Cactus Wren and Curve-billed Thrasher nests, April 1970, in cane
         cholla in lot 7 on Kleindale Road, Tucson, Arizona 172
18.2   1963 Curve-billed Thrasher breeding nests in Saguaro National
         Monument 176
18.3   1964 Curve-billed Thrasher breeding nests in Saguaro National
         Monument 177
18.4   1965 Curve-billed Thrasher breeding nests in Saguaro National
         Monument 177
18.5   1966 Curve-billed Thrasher breeding nests in Saguaro National
         Monument. 178
18.6   1967 Curve-billed Thrasher breeding nests in Saguaro National
         Monument 179
18.7   1968 Curve-billed Thrasher breeding nests in Saguaro National
         Monument. 179
18.8   Cactus Wren nests in Saguaro National Monument damaged or
         destroyed by Curve-billed Thrashers from August, 1962, to
         August, 1968 186

TABLES

3.1   Roosting nest locations. Saguaro National Monument. 22
3.2   Beginning of construction and date of occupancy of roosting nests.
        Kleindale Road area. 26
3.3   Orientation of Cactus Wren roosting nest entrance. (Near Santa Rita
        Mountains, summarized from Bailey, 1922.) 30
3.4   Orientation of Cactus Wren roosting nest entrance 30
7.1   Cactus Wren breeding nest locations, Saguaro National Monument,
        1963 to 1968. 63
7.2   Orientation of Cactus Wren breeding nest entrance on
        Kleindale Road, Tucson. 67
7.3   Orientation of Cactus Wren breeding nest entrance on Santa Rita
        Experimental Range, Arizona. 67

7.4   Orientation of Cactus Wren breeding nest entrance in Saguaro
       National Monument, Arizona. 67

7.5   Change in Cactus Wren nest locations in Saguaro
       National Monument, Arizona. 69

8.1   Kleindale Road. Date of first egg, with summary of temperature
       and rainfall. 75

8.2   Santa Rita Experimental Range. Date of first Cactus Wren
       egg laid (estimated). 79

8.3   Saguaro National Monument. Date of first Cactus Wren egg laid. 81

8.4   Saguaro National Monument. Climatic factors influencing start
       of Cactus Wren egg laying. 81

9.1   Kleindale Road area. Number of eggs per clutch of Cactus Wren. 85

9.2   Kleindale Road, 1939 to 1959. Number of Cactus Wren clutch sizes. 86

9.3   Saguaro National Monument. Cactus Wren clutches per territory. 87

9.4   Saguaro National Monument. Percentage of single, double, and
       triple Cactus Wren clutches. 87

9.5   Saguaro National Monument. Number of Cactus Wren clutch sizes. 89

9.6   Saguaro National Monument. Composition of first, second, and
       third clutches of Cactus Wren. 90

12.1   Hourly feeding rate at nest 19C on 19 April 1947. 104

13.1   Daily weights of Cactus Wren nestlings. 119

13.2   Mean daily weights of Cactus Wren nestlings. 121

14.1   Nest construction by first-year Cactus Wrens. 137

14.2   Nest construction by first-year Cactus Wrens of unknown age. 138

15.1   Hatching and fledging success of Cactus Wren on Kleindale Road. 142

15.2   Summary of Cactus Wren clutch failures, Kleindale Road. 143

15.3   Cactus Wren nesting success, Saguaro National Monument. 144

15.4   Percentage success of first, second, and third Cactus Wren clutches,
       Saguaro National Monument. 145

15.5   Annual successful Cactus Wren clutches, Saguaro National Monument. . . 145

15.6   Success of single, double, and triple clutches of Cactus Wren,
       Saguaro National Monument. 146

15.7   Cactus Wren breeding nest losses, Saguaro National Monument. 146

16.1   Percentage of Cactus Wren breeding population banded in
       Saguaro National Monument. 154

16.2   Loss of winter-banded Cactus Wrens in Saguaro National Monument. . . 156

16.3   Cactus Wren survival in Saguaro National Monument. 157

18.1   Loss of winter-banded Curve-billed Thrashers in
       Saguaro National Monument. 175

18.2   Date of first Curve-billed Thrasher egg, Saguaro National Monument. . . 181

18.3   Number of Curve-billed Thrasher clutches laid,
       Saguaro National Monument. 182

18.4   Size of Curve-billed Thrasher clutches laid in
       Saguaro National Monument. 182

18.5   Curve-billed Thrasher nesting success, Saguaro National Monument. . . 183

19.1   Other species of birds breeding in the Saguaro Monument area. 192

20.1   Comparative reproduction of Cactus Wrens and Curve-billed
       Thrashers, Saguaro National Monument, 1963 to 1968. 197

Foreword

Anders Anderson and his wife Anne, the junior author of this work, labored under real handicaps when they began their monograph investigations of the Cactus Wren in the 1930s. No particular literature to assist them was available; there was virtually no one to turn to for advice and inspiration. Furthermore, neither of them had a professional academic background in ornithology. For overcoming these obstacles — in fact, for not recognizing them as such — we salute them. They are truly pioneer Arizona ornithologists, members of a dwindling and different group.

When I first met Anders in Tucson in 1935, anyone in Arizona interested in birds was a rarity. Compare this with the era of the seventies when there is seemingly a bird-watcher behind every bush, and their number in Arizona runs into the hundreds. Compare also yesterday's bird literature with that of the 1970s. In 1935, the would-be Arizona bird student had little to go on except Chester Reed's *Bird Guide* and Florence Merriam Bailey's *Handbook of Birds of the Western United States* — both written shortly after 1900, and both lacking very much in information about the distribution of Arizona birds. Today, the market for excellent bird books that impart a wealth of information about birds in general and Arizona birds in particular is almost glutted.

Anders Anderson was born in Skutskär, Sweden, and emigrated to the United States with his parents in 1903, when he was four years old. The family settled in Bellingham, Washington, where Anders might have remained except for poor health which brought him to Arizona and Tucson in 1929 to recover. His vocation was that of electrician. Save for a hiatus at the bottom of the Great Depression in the 1930s, he was employed by the Tucson Gas and Electric Company and its predecessor organization. It was during his unemployed period that I made his acquaintance. He then had already completed his scholarly research that resulted in publication of an important paper, "The Arizona State

List Since 1914," in the *Condor* in 1934. Only someone well-versed in the field of ornithological taxonomy can appreciate the careful and meticulous effort that went into the appearance of this thoroughly scientific paper. For 30 years it stood as the only comprehensive addition to knowledge of the status and distribution of Arizona birds.

Anne Anderson, a native of Minneapolis, Minnesota, arrived in Tucson in 1926, three years before her future husband. A teacher in the Osakis, Minnesota, and Tucson, Arizona, schools, she and Anders were married in 1940. Anne, an enthusiastic and able amateur naturalist, spent years of weekends with her husband in their field of work on Cactus Wrens and Curve-billed Thrashers, as well as keeping a vigilant eye on bird happenings in their dooryard. It is safe to say that without her, this book could not have been written; Anders would be the first to say so.

There may be some scientists who might wish that the Andersons had carried out controlled experiments in the laboratory, to round out their work. Let it be said here that the lack of such does not detract from the present book in the least. The years of field work and the multi-hours of observation that went into this volume, coupled with a careful ecological approach to the life history of the Cactus Wren, make the Andersons' book unusually valuable and one of real stature in its field. And it was accomplished, in great part, on the authors' own resources of time and funds. A research volume is seldom so produced in modern times!

Arizona and Arizonans can well be proud of this book, whose subject is the Cactus Wren, the State bird.

GALE MONSON

Preface

This study of the Cactus Wren covers a period of about 38 years of intermittent effort. It began — after Dr. Walter P. Taylor of the United States Biological Survey in the early 1930s suggested that it would be a fascinating problem — as an endeavor to determine the actual behavior of a local pair of Cactus Wrens in the vicinity of our home near Rillito Creek, 6 miles northeast of downtown Tucson, Arizona. Later, starting in 1953, we devoted 3.5 years' work on a larger population of wrens on the Santa Rita Experimental Range, 35 miles south of Tucson. Finally, beginning in 1962, the senior author made a 6-year study of a population of Cactus Wrens and Curve-billed Thrashers in Saguaro National Monument, 14 miles east of Tucson. In addition to field work, we examined all important published references to the Cactus Wren.

The greater part of the Tucson and Santa Rita Experimental Range studies was published separately, from 1957 to 1965, in the *Condor,* journal of the Cooper Ornithological Society, under the titles "Life History of the Cactus Wren," and "The Cactus Wrens on the Santa Rita Experimental Range, Arizona." We wish to thank the editor of the *Condor* for permission to include this material here. We have rewritten and revised it, augmented it with new information, and rearranged it for a more — we hope — orderly, coherent presentation in book form.

We express our thanks to W. H. Behle, Herbert Friedmann, the late L. M. Huey, the late Seth Low, Margaret M. Nice, and A. L. Rand for aid in searching out bibliographic references, distribution records, and nesting areas; to the late Alden H. Miller and to Frank A. Pitelka for advice in preparation of the earlier manuscripts; to Gale Monson for his interest and encouragement in our seemingly endless task; to Joe T. Marshall, Jr. and Allan R. Phillips for nesting data; to the late J. J. Thornber, L. Benson, and C. T. Mason, Jr., for their frequent aid in identifying plant specimens; to J. E. Macdonald, J. R. Hastings, and

S. M. Alcorn for help in securing climatological reports. We are deeply grateful to Raymond Price, H. G. Reynolds, and S. C. Martin, who granted us permission to work on the Santa Rita Experimental Range, and to M. E. Fitch, W. E. Dyer, P. A. Judge, R. L. Giles, L. L. Gunzel, and H. R. Jones for their many courtesies at the Saguaro National Monument. A National Science Foundation Grant number GB-4003 provided assistance in the years 1965 to 1968.

We thank Harry L. and Ruth Crockett of Phoenix, Arizona, for their gracious permission to use their photograph of a Cactus Wren and its nestlings on the cover. We are grateful to the University of Arizona Press for the time and resources it made available for the production of this book, and for its interest and faith in our accomplishments.

ANDERS H. ANDERSON
ANNE ANDERSON

Introduction

The Cactus Wren (*Campylorhynchus brunneicapillus*) is a common resident of southwestern North American deserts, in thorn bush and arborescent cacti associations, from southern California, southern Nevada, southwestern Utah, western and southern Arizona, southern New Mexico, and central Texas southward to the tip of the Baja California peninsula, the west coast of Mexico south to Sinaloa, the east coast of Mexico to Tamaulipas, and on the Central Plateau to Michoacan, Mexico, and Hidalgo. Details of distribution of the seven geographic races are given in Appendix I, and chronology of the nomenclature in Appendix II.

On 16 March 1931, the regular session of the 10th Arizona State Legislature passed House Bill number 128, a part of which reads as follows: "Be it enacted by the Legislature of the State of Arizona. Section 1. The Cactus Wren, otherwise known as Coues' Cactus Wren or *Heleodytes brunneicapillus couesi* (Sharpe) shall be the state bird of Arizona."

In its physical characteristics the Cactus Wren belongs in the wren family Troglodytidae. To the ordinary observer, however, its large size, 7 to 8 inches (178 to 203 mm), its harsh, unmusical song, and its preference for a habitat with cacti, make it appear very unwren-like.

Its bill is almost as long as the head, and somewhat decurved at the end. In general, the top of the head and the back are brown, the feathers of the latter with narrow white streaks. The tail is mostly black, the outer feathers barred with white. A conspicuous white superciliary line is a good field mark. The underparts are white anteriorly, becoming rich buff posteriorly, the entire region spotted with black, the spots often much larger and denser on the chest. Juvenile birds can be distinguished by their lighter and smaller chest spots.

Our field observations have been particularly facilitated by the Cactus Wrens' nesting behavior. Both adult Cactus Wrens of a pair sleep in individual, separate roosting nests in their defended territory the entire year. Both share the labor of constructing their breeding nest. While the female incubates her eggs, the male usually begins work on one or more secondary nests. Sometimes such a nest is chosen for the next brood. It may also serve as temporary sleeping quarters for the female when her nest, crowded with nestlings, becomes uncomfortably cramped. Fledglings, too, may occupy these secondary nests.

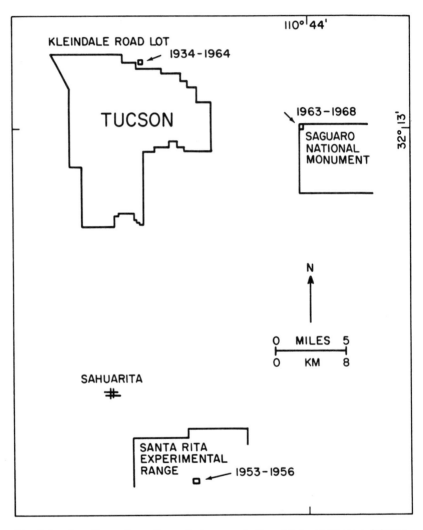

Fig. 1.1.   Locations of our three study areas in the vicinity of Tucson, Arizona.

I. Problems and Methods

We followed the procedure of first studying intensively the behavior of a single pair of birds and then extending our investigation to cover populations of several pairs (Fig. 1.1). Our field notes in the Kleindale Road block, based entirely on observations of living birds, cover a period of about 30 years. We collected no specimens, and tried to disturb the wrens as little as possible. Fortunately many of the observations could be made from the windows of our home, where we watched, unseen, the activities outside.

Progress was slow at first because we were working with a species in which the sexes are identical in appearance. Later, from 1939 to 1964, we colorbanded adults and nestlings, and obtained information of satisfactory reliability. Since we pursued the investigation entirely in our spare time, in mornings, evenings, and weekends, there were many delaying interruptions and digressions, resulting in a number of incomplete records. Nevertheless, we feel we gathered enough data to reveal clearly the behavior pattern of the Cactus Wren. It should be emphasized that our local population of wrens, living in a suburban environment and constantly harassed and disturbed by human activities, may not have been truly representative. Life in more open, natural surroundings may be somewhat different. In some respects it is perhaps harder. Predators are more numerous, and the food supply may fluctuate to a greater extent, than in the vicinity of human establishments. In the main, however, we found that the pattern of activity was the same on the open desert as that in our back lot.

We had no particular difficulty trapping wrens in accessible roosting nests after dark. We used a home-made trap of 0.25 inch (.63 cm) mesh hardware cloth, 6 x 6 x 12 inches (15.2 x 15.2 x 30.5 cm), closed at one end; the other end had a swinging door that could be closed the moment the wren entered. As soon as we thrust the trap over the nest entrance, the startled wren usually scrambled into the trap at once. This easy method, however, had the disadvantage of causing desertions of roosting nests. The wrens could not be returned successfully to their nests in darkness, but had to be kept indoors at night. In order to avoid this association by the wren of danger with its roosting nest, we substituted a small false-bottom or treadle-type trap on the ground, baited it with bits of bread and cotton, and confined our trapping efforts to daylight hours. We now encountered such competition from hungry House Sparrows (*Passer domesticus*), White-crowned Sparrows (*Zonotrichia leucophrys*), Curve-billed Thrashers (*Toxostoma curvirostre*), and even occasional Gila Woodpeckers (*Centurus uropygialis*), that we seldom captured the wrens at the desired time.

From 1939 to 1964 we banded 101 Cactus Wrens in the Kleindale Road area; 46 of these were adults or full-size immature birds and 55 were nestlings. All received numbered aluminum bands and one or two colored plastic bands to facilitate identification in the field. We found the colored bands could be recognized easily from a distance of at least 100 feet (30.5 m) with 8-power binoculars. For use in daily records we assigned to each wren the letter H, followed by a number indicating its chronologic entry in our banding log. When we determined the sex, we added the letter M for males and F for females: thus male and female became HM and HF. A bird not banded we call a noband. We numbered the cholla cacti used as nest sites arbitrarily; when the wrens built nests in them, each nest received the individual cholla number and a letter designating its order of construction thus: 6A, 14C.

We chose as our second research location, for a population study on a larger scale, the Santa Rita Experimental Range, 35 miles (56 km) south of Tucson, Arizona (Fig. 1.1). With a convenient barbed wire fence for the east boundary, and a diagonal northwest to southeast road for the north, we paced off and marked a trapezoidal plot of about 60 acres (24 ha). Since we were able to visit this somewhat distant area only on weekends, we attempted no banding, but devoted most of our time to the search for nests and to their periodic inspection. We began work on 10 January 1953, and stopped on 27 May 1956, visiting the location 88 times. The visits varied from 1 to 7 hours, averaging 3.79 hours. Later, we made three additional visits, one on 29 December 1956, one on 30 March 1958, and the last on 4 May 1958.

At the beginning of 1953 we located, tagged, and mapped all Cactus Wren roosting nests in the area. From then on, we endeavored to keep up with the new construction, recording on each visit progress and change in building activity, destruction or abandonment of nests, and presence or absence of Cactus Wrens at each nest. There are gaps in the record during the summer months, when vacations took us out of Arizona. This was especially the case in 1953. Several times, too, we found the area was inaccessible because of flooded roads from cloudburst type rains.

This study on the Santa Rita Experimental Range proved some-what unsatisfactory, chiefly because we did not band the birds. It left many questions unanswered. Losses and replacements, survival and dispersal of the young, and territorial and marital stability remained unknown. Furthermore, activities of the Curve-billed Thrasher, an apparently important competitor of the Cactus Wren for nest sites, and a competitor in part for food, deserved more attention. No one, apparently, had ever mentioned any intraspecific or interspecific conflicts of this nature in the case of the Cactus Wren, or had even commented on the disparity in numbers of individuals of the two species whenever they occur together in the same habitat.

All this finally led us, in the summer of 1962, to the selection of a third research site, this one in Saguaro National Monument (Fig. 1.1). It was closer to us, only a 14-mile (22.4 km) drive away, always accessible, and time now was available for trapping and banding. To some extent, the work was a duplication of our earlier research. However, it covered a period of 6 years, and the concurrent efforts directed at both Cactus Wrens and Curve-billed Thrashers produced a better understanding of the fortunes and fluctuations of the two populations. We feel that too little of living biological research is repeated. Researchers usually believe it is more productive to go on to something new. The recent study of the northwest coast Song Sparrows (*Melospiza melodia*) by Tompa (1962, 1963), which followed the early, monumental work of Nice (1937) on an Ohio population, is an excellent example of the value of duplicating research.

The fenced, northwest corner of Saguaro National Monument on Freeman Road provided two easily surveyed sides from which to lay out a square of 49 acres (19.6 ha). From this 0.0 corner, the north side received stakes at acre (.4 ha) intervals, that is, at every 208.7 feet (62.6 m), lettered from A to G, and the west side received stakes numbered from 1 to 7. All interior acre stakes (paced) had combinations of the two to provide coordinates for mapping nest locations.

Weekly visits were made to the research area until the end of 1963, followed by an average of eight visits per month in 1964, ten per month in 1965, and then about 12 per month until the study ended in July 1968. Summer visits were short, usually only 3 or 4 hours; in the cooler months they averaged as much as 7 hours each visit. After the initial survey of nests in the late summer of 1962, the entire area was checked once a month.

Fifteen small Bailey treadle-type traps, baited with bread or potatoes and a few pieces of cotton, were used in capturing 81 adult Cactus Wrens and 29 adult Curve-billed Thrashers in the course of the 6 years. To minimize disturbance, trapping was discontinued during the breeding season. Out of an estimated grand total of 538 Cactus Wren nestlings produced on the area, 389 (72.3 percent) were banded; all 108 of the

Curve-billed Thrasher nestlings were banded. There were no nest desertions. Some of the Cactus Wren nests were inaccessible, a few because
of their extreme height in saguaros, others because they were so tightly
wedged between the thick branches of the saguaros that the hand could
not be inserted to remove the nestlings. As in the earlier studies, field
identification was aided by placing the aluminum band on the right leg
of the adults and on the left leg of nestlings. All received various combinations of colored plastic bands. After 6 years it appeared that the number of possible, useful, color combinations for the Cactus Wrens was almost
exhausted. To continue banding would necessitate duplication of colors
and result in almost certain increased errors in field identification. To
avoid confusion with the Kleindale Road records, the Saguaro Monument
wrens were given the prefix C instead of H. Curve-billed Thrashers were
designated by the letter T.

The observation, recognition, and tracing of the color-banded birds
in the 49-acre (19.6 ha) plot became a most time-consuming task, even
with the aid of the indispensible 15- to 60-power "zoom" telescope. (The
narrow, two-color bands were often difficult to distinguish with binoculars.) Evidence of dispersion or relocation usually remained hidden
until the males proclaimed their new residence by means of their songs
in the early spring. The approximate territorial boundaries could then
be determined by mapping their singing stations, and, especially in the case
of the Cactus Wrens, by noting their border conflicts. Allocation of females
became relatively easy only when nest building started.

At first we determined the sex of our wrens from their behaviour
in the course of the display-growl threat; we seldom witnessed copulation.
Later we observed that females rarely sing, and they alone incubated
the eggs.

2. The Habitat

Kleindale Road location. Our home, built on lot 7 in a block of ten commercial-shaped acres, became the center of our first studies (Fig. 2.1). About 400 feet (121.9 m) to the north of this block, a rather dense growth of mesquite (*Prosopis juliflora*) and catclaw (*Acacia greggii*) bordered the steep bank of Rillito Creek. In the creek's sandy bed grew a few cottonwood (*Populus fremontii*) trees and willows (*Salix* sp.). The elevation is 2,400 feet (731.5 m) above sea level; the land slopes gently down to the creek. In 1939, five of the acre tracts on Kleindale Road and one on Greenlee Road contained small residences and various auxiliary buildings such as chicken houses and garages. The remaining 4 acres were unoccupied. Creosote bush (*Larrea divaricata*), widespread indicator of the Lower Sonoran Life Zone, predominated not only in the study area, but also for considerable distances to the east, south, and west. Its density varied greatly. Our lot, which we purposely left in its original condition as much as possible, contained at least 350 shrubs of this species, ranging in height from 1 to 6 feet (.3 to 1.8 m). Across the southwest corner of the tract, running from southeast to northwest, parts of an old, shallow, abandoned irrigation ditch provided a foothold for a ragged growth of low mesquite and catclaw. A few cultivated trees, athel (*Tamarix aphylla*) and China-berry (*Melia azederach*), had attained adult height along the front of the residences on Kleindale Road. Here and there cholla cacti emerged in the almost uniform expanse of creosote bush. Our lot also supported a maximum number of these — about 20 plants of sufficient size to furnish shelter and nesting sites for Cactus Wrens. Cane cholla (*Opuntia spinosior*) was the common species, with a few jumping cholla (*O. fulgida*) and staghorn cholla (*O. versicolor*) interspersed. Other shrubs — desert broom (*Baccharis sarothroides*), desert thorn (*Lycium berlandieri*), and Mormon tea (*Ephedra*

[7]

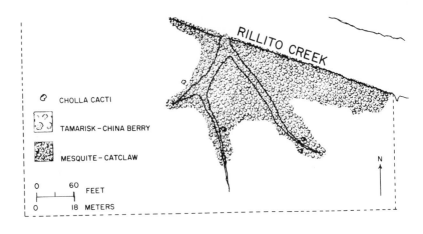

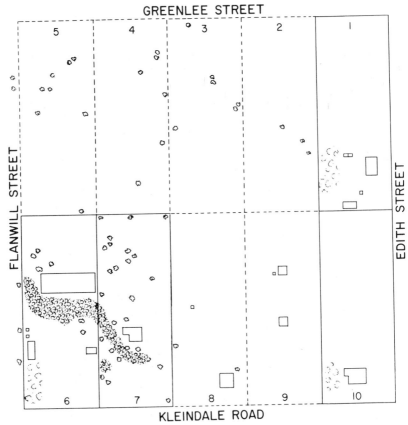

Fig. 2.1. The Kleindale Road, Tucson, Arizona, study area in 1941.
Solid lines indicate fences; dashed lines, lot boundaries.
Creosote bush association covered all unshaded areas except streets.

trifurca) — were few in number and of little significance in relation to shelter or nesting facilities for birds (Fig. 2.2).

For the most part, the soil was a bare, brown, sandy loam. If winter rains were abundant, numerous ephemeral plants appeared in early spring, rapidly carpeting the ground with small flowers. The most conspicuous were filaree (*Erodium cicutarium*), which sometimes emerged in December, and bladder-pod (*Lesquerella gordoni*). Other common spring ephemerals were: the grasses *Festuca octoflora* and *Schismus barbatus;* the mustards *Lepidium lasiocarpum, Sisymbrium irio, Descurainia pinnata,* and *D. sophia; Astragalus nuttallianus; Erodium texanum; Bowlesia incana; Phacelia crenulata; Pectocarya platycarpa* and *P. recurvata; Lappula redowskii; Cryptantha angustifolia* and *C. barbigera; Amsinckia intermedia;* the plantains *Plantago insularis* and *P. purshii; Stylocline micropoides;* and *Evax multicaulis.* All these completed their cycles of flower and fruit by mid-May, and then dried up in the increasing daily temperatures.

After the summer rains, which usually began in July, another group of ephemerals appeared, less numerous in species, but often equally dense and conspicuous. The most abundant were: six-weeks grama (*Bouteloua barbata*) and needle grama (*B. aristidoides*); *Eriogonum deflexum* and *E. trichopes; Boerhaavia caribaea, B. spicata,* and *B. wrightii; Tribulus terrestris;* and *Kallstroemia grandiflora* and *K. parviflora.*

Ordinarily no water was available in the 10 acres (4 ha), other than that from dripping faucets or bird bath pools constructed by interested residents. Rillito Creek, normally a dry bed of sand, carried water only after summer cloudbursts in the nearby mountains, or after unusually heavy winter rainfall, and then only for a relatively few days. The brief, sometimes torrential, summer rains often scoured away portions of the top soil in the vicinity, for the plant cover was seldom heavy enough to prevent erosion.

The appearance of the dominant creosote bush south of Rillito Creek changed very little from 1930 to 1957. The growth of this shrub is normally slow, and its age is difficult to determine. According to Shreve (1951: 157), its age probably "greatly exceeds 100 years." Many creosote bushes in the area in 1930 were old, with dead, dry, drooping outer branches. Small seedlings were very scarce; they had evidently been crowded out long ago. If there ever had been successional stages leading to this climax community, the remnants of these stages were gone. Although portions of an abandoned irrigation ditch were present in 1930, there is no evidence that water had been used in the vicinity of Kleindale Road. The ditch merely crossed the area. It seems safe to assume that the creosote bush association had been here for at least 50 years; it probably was here when the first Spaniards arrived in the sixteenth century.

A small invasion of cholla cacti into the creosote bush area evidently was relatively recent. In 1930, these plants were widely spaced. They occupied a strip of land approximately 0.5 mile long by 0.1 mile wide

Fig. 2.2. The Kleindale Road, Tucson, Arizona, study area,
looking north from Lot 7 in 1938.

(.8 by .16 km), immediately south of the Rillito streamside fringe of
mesquite and catclaw. There was no tendency toward colonialism, such
as occurs in mature stands. No large dead or dying chollas, with basal
circlets of new, young seedlings were present. The invasion could have
begun with introduction of cattle into the area by early settlers, as cattle
are well-known disseminators of cholla cacti. The spiny joints of jumping
chollas are easily detached when an animal brushes against the shrub.
These joints cling to the cow's hide and are often transported long distances before falling to the ground. There they take root readily. Some
of the cane chollas may have been introduced by rodents. Round-tailed
Ground Squirrels (*Citellus tereticaudus*) and Antelope Ground Squirrels
(*C. harrisii*) were common in the vicinity in 1930. Both species have
been observed carrying cholla fruits. The nearest extensive cholla stand
is on the north side of the Rillito, at the foot of the bajadas of the Santa
Catalina Mountains.

Cactus Wrens do not inhabit a pure creosote bush association, for
it has no nesting sites. Evidently the wrens extended their range southward across the Rillito at the time the cholla cacti became large enough
to live in. Judging from the rate of cholla growth in lot 7, this occupation
probably did not occur before 1915.

The fast-growing city of Tucson produced many changes in the
Kleindale Road area in the course of our 30-year study (Fig. 2.3). While
we tried to preserve as much as possible of the original vegetation on
our lot, we had no control over the lots of our neighbors. A sand and
gravel company took over the south bank of Rillito Creek, removed the

vegetation, and sold the top-soil. The creosote bushes south of Kleindale Road were bulldozed away to make room for 25 small houses. Most of the chollas on the tract, with the exception of those on our lot, gradually disappeared, for the new residents found their spiny joints dangerous to children and pets. Lots 2, 3, 4, and 5 along Greenlee Road became a riding stable. Ornamental plantings increased. Numerous electric and telephone poles framed the area; the grading of new streets changed and diverted the natural drainage channels.

Santa Rita Experimental Range. In our search for a larger study area, we selected a 60-acre (24 ha) plot in Pasture 5, 9 miles (14.4 km) from Sahuarita on the Sahuarita-Helvetia road. The Range seemed to offer several advantages. It was closed to hunting; its interior roads were too narrow and rough to attract many visitors; and temperature and precipitation were being recorded at a number of stations. The only serious disturbances to its value as a natural area were predator control and the introduction of cattle. The cattle, however, were limited under the supervision of the United States Department of Agriculture to from six to eight head per square mile (1.6 km²).

The elevation is about 3,300 feet (1,005 m); the land slopes westward. At the middle of the east boundary fence of the pasture, a large iron pan, fed by a larger, concrete reservoir, designated as the "North Rim" on the Forest Service map, provided water for cattle assigned to this portion of Pasture 5. Most of the soil was a coarse sandy loam. Generally the vegetation of this research plot was the "cholla meadow" of Brandt (1951: 57), divided into roughly triangular patterns by strips

Fig. 2.3.   The Kleindale Road, Tucson, Arizona, study area, looking south through Lot 7 in 1956.

Fig. 2.4.   Santa Rita Experimental Range, 35 miles (56 km)
south of Tucson, Arizona, 19 December, 1954: cholla association.

of other associations. Under the Shreve (1951: 40) classification it would
fall into the upper border of the Cercidium-Opuntia region, Arizona Upland, Upper Bajada, with ribbons of Streamway vegetation. There was
an understory of scattered burroweed (*Haplopappus tenuisectus*). Had
drainage channels been absent, it is probable that the entire area would
have been dominated by cholla cacti of two species, *Opuntia fulgida,*
including a less spiny variety, *mammillata,* and *O. spinosior.* Usually each
species grew in almost pure stands, often very dense in the case of *O. fulgida.* Their heights ranged from 3 to 6 feet (.9 to 1.8 m). Where intermingling of the two species occurred, the spacing between plants was
wider (Fig. 2.4). Following the chollas in abundance was prickly pear,
Opuntia engelmannii. It was generally distributed, but was seldom of sufficient height to be conspicuous.

The uniformity of the cholla association was broken by several
normally dry, sandy washes and their tributaries, running irregularly
from southeast to northwest across the south half of the plot. They were
bordered by a distinct wash association (Fig. 2.5), consisting of a fringe
of mesquite, blue palo verde (*Cercidium floridum*), and catclaw; the
first two grew up to 20 feet (6 m) high and the last one to 15 feet (4.6 m).
Impenetrable clumps of desert hackberry (*Celtis pallida*), up to 9 feet
(2.7 m) in height, lined the edges of some of the channels, forming
barriers of considerable length. Between the washes many scattered individuals of these four species had found a foothold. To a lesser degree,
this invasion extended even into the pure cholla association on either
side. Other less common perennials were desert honeysuckle (*Anisacanthus thurberi*), Mormon tea, ocotillo (*Fouquieria splendens*), gray-thorn
(*Condalia lycioides*), fairy duster (*Calliandra eriophylla*), and *Baccharis
brachyphylla.*

Creosote bush, so abundant on the Kleindale Road area, was absent on the Range plot. Several species of perennial grasses, tanglehead (*Heteropogon contortus*), gramas (*Bouteloua* spp.), three-awn (*Aristida* sp.), and cottongrass (*Trichachne californica*), were present, chiefly along the borders of the washes. Mesquite-grass (*Muhlenbergia porteri*) was more widespread, occurring under the protective cover of various shrubs. The winter ephemerals were less numerous in species than in the Kleindale Road tract. *Eriogonums, Cryptanthas,* and *Erodium* were common. Among the summer ephemerals, needle grama was by far the most abundant. Grasses were particularly conspicuous after the 1954 summer rains.

Saguaro National Monument. This, too, is classified by Shreve (1951: 40) as being in the Arizona Upland, Upper Bajada, but it differed from the Santa Rita study area not only in the density of the stem-succulent species, but also in their composition. Its aspect was that of a saguaro-palo verde-cactus association. Brandt (1951: 36) named the association a Giant Cactus Park. The study area, located in section 17 of the eastern portion of the Monument (Lat. 32° 13', Long. 110° 44'), has an elevation of about 2,840 feet (852 m), with a gradual slope from south to north of from 2 to 3 percent. Seven miles (11.2 km) to the north

Fig. 2.5.   Santa Rita Experimental Range, 35 miles (56 km) south of Tucson, Arizona, 19 December, 1954: wash association.

rose the Santa Catalina Mountains, and 3 miles (4.8 km) to the east, the Rincon Mountains.

The several shallow swales, which began at the south boundary of the 49-acre area, widened to 100 feet (30.5 m) or more and deepened to 6 to 9 feet (1.8 to 2.7 m) on their courses northward. At their centers narrow, sandy washes appeared, 4 to 9 feet (1.2 to 2.7 m) wide, their minor tributaries joining them in typical tree-branch pattern; nowhere had these washes progressed to the point of becoming deep-cut arroyos. The soil in the open, flat areas was a light brown sand, intermixed with angular gravel particles, often only 0.25 to 1.0 inch (0.63 to 2.5 cm) in width. On the slopes, scattered irregular stones, 2 to 5 inches (5 to 12.7 cm) across, were common.

Two large plants dominated the landscape, the saguaro (*Carnegiea gigantea*) growing occasionally up to 32 feet (9.8 m) in height, but usually less, and the more numerous foothill palo verde (*Cercidium microphyllum*), 10 to 20 feet (3.1 to 6.2 m) in height, with low, broad crowns (Figs. 2.6, 2.7). Saguaros ranged from none to 39 per acre (0.4 ha) quadrat; the average was about 10. Only two quadrats had none. Palo verdes ranged from 2 to 47 per quadrat, the average 22. They were heavily parasitized by leafless mistletoe (*Phoradendron californicum*). Some of the larger trees had died. A few scattered mesquites and still fewer blue palo verdes grew chiefly in the washes. White-thorn (*Acacia constricta*) grew abundantly in and along the washes, sometimes forming a continuous border, now and then reaching a height of 9 or more feet (2.7 m) and spreading their diameter to 12 feet (3.7 m). Catclaw appeared less commonly. Desert hackberry, 4 to 10 feet (1.2 to 3.1 m) high, and Mexican crucillo (*Condalia spathulata*), 4 to 6 feet (1.2 to 1.8 m) high, joined in forming extensive, impenetrable thickets in and near the main washes. Gray-thorn and ocotillo were distributed irregularly in the research area. Creosote bush overran several quadrats along Freeman Road, but decreased in density rapidly eastward, to reappear in greater numbers at the east border. All the above plants, except creosote bush, had thorns, spines, or spine-tipped branches.

Five species of cholla cacti were irregularly distributed among the larger vegetation. Of these, only jumping cholla and its more common and less spiny variety *mammillata,* and staghorn cholla, occurred in sufficient numbers and size to be important as nest sites. Together, these two species averaged about 15 plants above 2 feet (0.6 m) in height per quadrat. They were absent in three quadrats; the range was none to 43. Although staghorn cholla was commonly more scattered, jumping cholla sometimes occurred in groves (Fig. 2.8) of 10 to 15 individuals with dense crowns, in heights up to 7 to 8 feet (2.1 to 2.4 m), with the average about 5 feet (1.5 m). Many larger, isolated plants had shed their lower branches; others were dying, or were already dead, with blackened joints. The staghorn cholla, too, had apparently attained its maximum growth. A considerable number were standing dry and dead, a few fallen. At first

Fig. 2.6.   Saguaro National Monument east of Tucson, Arizona. Looking
north into Quadrat E2, 11 December, 1966, the Santa Catalina Mountains
in the background. The columnar cacti are saguaros. Most of the low,
arborescent cacti are jumping chollas; the leafless trees are foothill palo verdes.

Fig. 2.7.   Saguaro National Monument east of Tucson, Arizona.
The saguaro grove in Quadrat C4, 11 June, 1968.

Fig. 2.8.   Saguaro National Monument east of Tucson, Arizona.
Looking north into the C3 cholla grove, 11 December, 1966.

glance these species appeared to be declining, but on closer inspection one found numerous small replacements 2 to 4 inches (5 to 10 cm) high in the vicinity of the parent plant. Woodrats (*Neotoma*) gathered the abundant litter from these chollas to cover their dens.

All this perennial vegetation, despite its size and numbers, covered only part of the area. Nowhere was the arborescent canopy completely closed. The numerous remaining open spaces provided room for 15 to 20 scattered herbaceous or shrubby perennial species 1 to 3 feet (0.3 to 0.9 m) in height, burroweed being the most numerous. Broad mats of two species of prickly pear (*Opuntia engelmannii* and *O. phaeacantha*) and clumps of perennial grasses were common. After good winter rains, at least 30 species of winter ephemerals appeared and occupied the intervening open spaces. Most abundant were *Pectocarya* sp., *Plantago* sp., *Lotus* sp., the introduced *Erodium,* and several species of mustards. As in the other two research areas, these plants became dry and shriveled by May. After July rains, sixweeks grama grasses (*Bouteloua* spp.) predominated.

Compared with the Monument area, the Kleindale Road locality appeared depauperate, for it lacked the saguaro and palo verde. Common to all three areas were the cholla cacti, the apparent requisite in the life of the Cactus Wren. Saguaros were absent on the Santa Rita Experimental Range plot, and blue palo verde, not foothill palo verde, was the principal tree. Cattle were excluded from the Monument.

Climate: The climate of the Tucson area is characterized by a long hot season, beginning in April and ending in October (U. S. Weather Bureau, 1966). "From May through September, maximum temperatures above 90F [32.2C] are the rule, with the mean maximum occasionally exceeding 100F [37.8C] in July. Under usual conditions, the diurnal temperature range is large, averaging almost 30F [16.7C], although it may exceed 40F [22.2C]. Clear skies or very thin high clouds permit intense surface heating during the day and active radiational cooling at night, a process enhanced by the characteristic atmospheric dryness. The average growing season in the Tucson area approximates 250 days.

"Most of the year's precipitation falls in two main periods: more than 50% of it, usually between July 1st and September 15th, when scattered convective or orographic showers and thunderstorms occur. . . . The period providing the secondary precipitation maximum, December through March, sees more general, prolonged rainstorms . . ." Annual precipitation at Tucson from 1930 to 1965 varied from a low of 5.34 inches (135.6 mm) in 1953 to a high of 17.99 inches (456.9 mm) in 1964; the mean was 11.17 inches (283.7 mm).

Relative humidity varies with the season and the time of day. During the hot season it may fall below 10 percent, even sometimes below 5 percent.

"Surface winds are generally light, with no important seasonal changes in either velocities or prevailing direction . . . Wind velocities and direction are influenced to an important extent by the surrounding mountains, as well as by the general slope of the terrain. With weak pressure gradients, local winds tend to be in the SE quadrant during the night and early morning hours, veering to NW during the day. Highest velocities usually occur with winds from the SW and E" (U. S. Weather Bureau, 1966).

In western Arizona and along the northern periphery of the Cactus Wren's range in southern Nevada and southwestern Utah, the precipitation is much lower, and in southwestern Texas, near the Gulf of Mexico, it is twice as great as that of the Tucson region. Differences in temperatures also occur, and the changes in the vegetation are striking. So far, no extensive investigations of Cactus Wren behavior in these areas have been published.

3. Winter Behavior and Roosting Nests

Cactus Wrens appear to be strictly resident. The adult wrens we banded remained with us throughout the winter with extremely slight shifts in territory. We never observed any increase in the wren population during the early spring months, as would have occurred had there been an influx of migrants from the south. The population peak, which was attained in late summer, was produced entirely by the addition of immature birds that had been raised in the vicinity. These intermingled with the adults and tentatively probed into adjacent territories. The population gradually decreased as the immature birds left or vanished, reaching its lowest point in the winter. Adult wrens which suddenly disappeared usually were replaced almost immediately by new birds. Evidently unmated individuals, in nearby but less desirable territories, moved whenever possible into more favorable locations.

At Tucson the period of what might be termed routine winter activity was very brief. November could be called the dormant, stable month. Each year's cycle actually began in December. The basic "population" of one pair of Cactus Wrens occupied approximately the same territory it had used in the course of the spring and summer months for breeding purposes. In addition, this pair usually tolerated, in loose attachment, some immature birds of the preceding nesting period, chiefly the surviving members of the pair's own offspring, which were now in adult plumage, and an occasional outsider. Often the group foraged as a unit, without apparent antagonism between any of its members. At times, some of the wrens would venture north to the bank of the Rillito, or would trespass into land occupied by other wrens. Although flocking behavior has been reported, our local population was never augmented by the addition of wrens from neighboring territories in search of food.

Howell (1916: 213-214) saw flocks of from six to 30 or more Cactus Wrens going through tops of cottonwoods along the Rillito in cold weather. It is not surprising that they would visit the cottonwoods, for these trees are adjacent to the cactus-covered bajadas on the north bank of the Rillito, and would represent only a small extension of territory. It is the number of wrens which is exceptional. We never have observed Cactus Wrens in such large numbers anywhere in Arizona, nor have we been able to find another reference in the literature to such extreme flocking behavior. Conceivably the combination of a successful breeding season with the survival of most of the young birds of three or four broods could produce such an extraordinarily large family group. Otherwise it would require the supposition that the occupants of at least half a dozen adjacent territories combine with their offspring into a compact group for the purpose of foraging.

Howell did not report the extent of the cottonwood foraging expedition. It seems probable that it was the case of a sudden, abundant, concentrated food supply in the center of a considerable population of wrens. The birds of the surrounding area, within sight of each other, would be attracted to this center radially as though by centripetal force, and they would disperse to their respective sectors in the circle as soon as satisfied. An observation by Dr. R. B. Streets of the University of Arizona (personal communication) would seem to support this view. On 4 January 1953, he saw "large numbers" of Cactus Wrens in the tops of date palms in a small grove at a ranch near Tanque Verde Creek east of Tucson. On later visits by us, we found no wrens in the palms, but saw them in the nearby desert in their usual habitat.

The abandonment, even temporarily, of a definite territory – and the Cactus Wrens in our area were strictly territorial in their habits – for communal feeding seems illogical. Nevertheless, it must be considered, for there were instances in our later work when the absence of wrens on the Range and the Monument could not be explained in any other way. On none of the visits to the Monument in the winter could all the wrens of the previous year's territories be found. In the late afternoon of 20 January 1964, almost the entire 49 acres appeared vacant. Later in the day some of the wrens returned. On 8 December 1965, at least seven wrens moved northward, traversing several territories. They fed on the ground among the low tufts of grass, each bird apparently paying no attention to the others, except that it continued to remain in the loose formation. This temporary breakdown of territorial boundaries probably lasts only a few hours at most. By sunset, the wrens must return to the vicinity of their roosting nests in their respective territories.

The Kleindale Road feeding area must have included at least 15 acres (6 ha) during the winter months. It is doubtful if Cactus Wrens visited all this area regularly each day. Frequently the group of birds stayed in the vicinity of the houses where food was more plentiful.

Our Cactus Wrens required a covered roosting nest in all months of the year. In the Tucson area, each adult wren occupied nightly its own nest. At no time did we find more than one wren in a winter roosting nest. Any available nest was used as long as it was habitable. Usually by the end of the breeding season some of the older nests had deteriorated; others had been relinquished to immature birds. The adults selected new sites and built again. Ordinarily under more normal, or perhaps ideal, conditions, these roosting nests probably would have been used throughout the winter. Here, however, because of frequent nest destruction by Curve-billed Thrashers, the Cactus Wrens were engaged in building during the entire winter.

The location of roosting nests naturally depended upon the availability of cholla cacti. These cacti were most numerous in the western portion of the Kleindale Road 10 acres. The usual territorial area of a pair of wrens here was about 10 acres, with irregular additions of 5 adjacent acres. The center of activity, however, lay not in the center of the tract, but far to the western edge. As shown in Fig. 3.1, which indicates the locations of the roosting nests during the months of November and December for the years 1938, 1939, 1943, 1944, 1946, and 1947, lots 6 and 7 were preferred by the resident pair.

The roosting nests of the Kleindale Road pair were seldom far apart. In three of the years, they were in the same cholla. The maximum separation of 140 feet (42 m) occurred in 1938. Distances to other members of the winter group varied considerably; some were as close as 100 feet (30.5 m), others much farther. A few nests were never located, suggesting the probability that some of the wrens had strayed in temporarily from the area to the east. The daily variation in the local winter population of from three to eight birds would tend to confirm this. There were not enough roosting nests in the territory to house the entire population.

The Cactus Wrens used the cholla cacti almost exclusively for nest sites. Only when the wrens were hard pressed by competition did they use other places. These exceptions were catclaw bushes, an ornamental pyracantha, a House Sparrow roost under the eaves of a garage, and a House Sparrow nest box in the back lot, of which the owners were dispossessed by wrens. Nest heights, of course, were limited by the heights of the chollas, the maximum being 8 feet (2.4 m). Because of its open growth, cane cholla seldom offered a site lower than 3 feet (0.9 m). Jumping chollas had lower, denser crowns, yet no nests were placed below 3 feet. The average height, determined partly by the configuration of the cholla crown, and partly by the choice of the wren, was from 4 to 5 feet (1.2 to 1.5 m).

On the Santa Rita Experimental Range, Cactus Wrens constructed all their roosting and breeding nests in cholla cacti, despite the availability of densely-branched palo verde and mesquite trees with apparently suitable nest locations in their mistletoe growths. Thirty-two years earlier, Bailey (1922: 163-168), working at an elevation of approximately 4,000 feet (1,200 m) and about 7 miles (11.2 km) south of our Santa Rita Range plot, reported that in her 53-acre (21.5 ha) study area none of

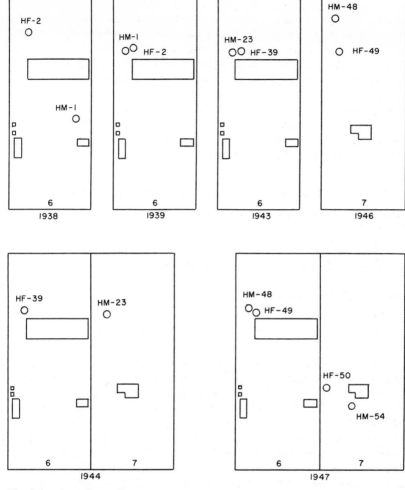

Fig. 3.1. Locations of roosting nests of a pair of Cactus Wrens
in Lots 6 and 7 on Kleindale Road, Tucson, Arizona, in
November and December of 1938, 1939, 1943, 1944, 1946, and 1947.
Circle with wren's number indicates nest.

the roosting nests were placed in cholla cacti. She found six nests in
catclaw, 15 in clusters of mistletoe in catclaw, 15 in *Zizyphus* (=*Condalia
lycioides*), and 1 in mistletoe in a mesquite tree. It is not clear whether
cholla cacti were present in this area, but "cholla cactus flats of the lower
terraces" are mentioned (p. 163). Cacti are not numerous above 4,000
feet on the Range, and this elevation is also close to the upper limit of
the Cactus Wren's distribution in Arizona.

The sites of 528 roosting nests of the Cactus Wren in Saguaro National Monument are listed in Table 3.1. One nest in an unusually large, robust pencil cholla, *Opuntia arbuscula,* is included under *O. spinosior.* The total includes many nests which were begun in the autumn and abandoned for various reasons before completion. The dense clumps of mistletoe in palo verde trees offered good concealment. Roosting nests in saguaros rested in crotches between the branches; a few were in large, callused holes in the trunk. The upright, ribbed cages of dead, dried saguaro stumps also offered space and concealment for nests. The differences in totals for chollas reflect the number of suitable plants available, and not a preference for any particular species.

Table 3.1. Roosting nest locations. Saguaro National Monument.

Opuntia fulgida and O. fulgida var. mammillata	344
Opuntia versicolor	65
Opuntia spinosior	12
Cercidium microphyllum	19
Phoradendron californicum in Cercidium microphyllum	18
Carnegiea gigantea	50
Carnegiea gigantea — stump	20
TOTAL	**528**

There is no essential difference between the roosting nest and the breeding nest. In fact, they are sometimes interchangeable. The Cactus Wren's nest is probably the best known part of this interesting bird's life history (Bailey, 1922: 163-168; Woods in Bent, 1948: 220-223; Brandt, 1951: 679-680), for the nests are too conspicuous to overlook or neglect. Many variations in structure have been observed, but none was noted which differed from the basic pouch design with an entrance at one end. The "standard" or "normal" nest is one about 12 inches (30.5 cm) long, sloping downward from the entrance at a 30 degree angle. It has an entrance roughly 1.5 inches (3.8 cm) in diameter, and a well-defined cylindrical vestibule of the same diameter, but of varying length, leading to a nest cavity, into which it drops abruptly. The cavity may be 3 inches (7.5 cm) in diameter. In profile the exterior is pouch-shaped; the interior is formed like a retort.

Variations in the position and shape of the nest undoubtedly sometimes are caused by the wren's inability to choose the proper site (Fig. 3.2). Such inability is not necessarily due to inexperience, for adults may have the same difficulties in choosing a nest site as immature birds. Nest sites are never exactly alike. The floor of the nest cavity requires a support of cactus joints or twigs. If the place selected is of insufficient length, the vestibule will be shortened or even absent. The nest may be little more than a wide-mouthed tumbler set on edge. Occasionally this prob-

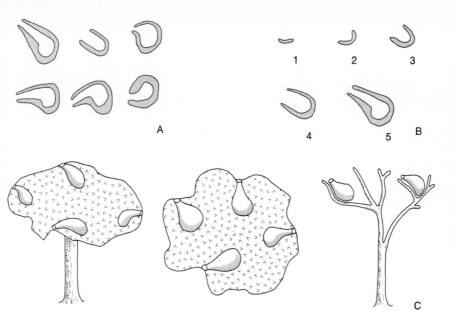

Fig. 3.2. A. Common variations in the shape of Cactus Wrens' nests. B. Stages in the construction of a roosting nest. Occupation usually begins at stage 3. C. Common locations of roosting nests in cholla cacti; left, **Opuntia fulgida;** center, **O. fulgida,** top view; right, **O. spinosior.** Nest entrance usually faces outward. The number of nests in a cholla varies, there being seldom more than two usable nests.

lem is solved by pushing more and more material into the cavity, forcing the cavity backward and downward until it becomes suspended from the vestibule without under support. This situation can lead to changes in the slope of the vestibule. A longer floor results in a longer vestibule; sometimes this vestibule is 12 to 15 inches (30.5 to 38.1 cm) long before it ends in a flaring cone of grass stems. A "doorstep" of some sort is always necessary, for the entrance is too small to admit a flying bird, and the funnel of grasses is too weak to support a wren's weight. This "doorstep" is usually a twig or a branch growing below or at the side of the entrance.

The general form of the nest is apparently a reflection of inherited behavior, but the material used in construction depends upon what is available in the vicinity of the particular site. On the Santa Rita Experimental Range, and in other parts of the better-watered, eastern Arizona desert, the nests were constructed of dried grasses, with a weak skeleton framework of coarser grasses and such intricately branched fall ephemerals as the *Boerhaavias*. Bailey (166-167) lists 24 species of plants, most of them annuals, used in nest construction on the study area at the north base of the Santa Rita Mountains. Farther westward, in the Organ Pipe Cactus National Monument in southwestern Arizona, where grasses are often less abundant, we found *Eriogonum deflexum* used almost exclusively for the exterior walls of the nests. The dark, reddish-brown color of

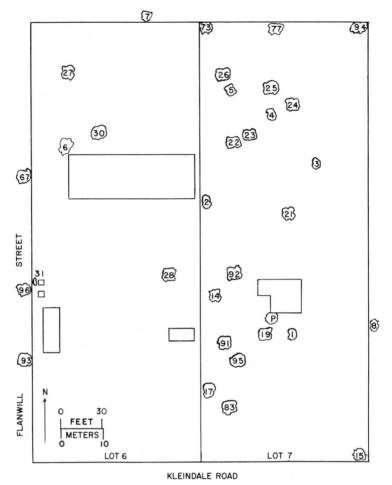

Fig. 3.3.   Location of numbered cholla cacti in Lots 6 and 7 along Kleindale Road, Tucson, Arizona.

these nests was in striking contrast to the pale straw hue of the eastern nests. Along the arid, western slope of the Beaver Dam Mountains in southwestern Utah, Joshua trees (*Yucca brevifolia*) provide a supply of dry, shredded fibers for nest material. The few available grasses are used in the interior lining. When possible, the nest cavity is carpeted with feathers. Plant down seldom is used.

When civilization creeps into the wren's domain, an abrupt change takes place in the nest materials used. Almost anything imaginable is gathered up and fashioned into a nest. Bits of newspaper, tissue paper, cotton, string, rope, rags, fur, lint, and, above all, chicken feathers, replace the native materials.

We found it difficult to observe the act of selecting a nest site. In most cases we did not notice any preliminary activity, for our attention was attracted only after construction began. (See Fig. 3.3 for locations of numbered cholla cacti in lots 6 and 7 along Kleindale Road, Tucson, Arizona.) Indecision or uncertainty must occur, however, because it is not unusual to discover small bits of nest material that have been placed in various chollas and then abandoned. An extreme example is that of H-35, a wren which had lost two roosting nests in the fall of 1941, the last on 6 December. On 7 December, at 0950, this wren inspected two damaged nests in cholla 23. It pulled a straw from nest 23A and carried it around to nest 23B in the same cholla. Evidently this latter nest was not satisfactory, for it now began gathering material on the ground and placing it in cholla 3. Soon it was back to nest 23B, trying to straighten out the disarray of grasses and then adding new material. Again it left. This time it did not return to work until 1130. Still not satisfied, it moved to cholla 22 and started another foundation. This work was interrupted when a Curve-billed Thrasher climbed the cholla and took up a position on the floor of the nest. H-35 eyed the thrasher a moment; then it flew to cholla 3, where it apparently looked for another site. Undecided again, it flew to cholla 14 and inspected an old nest remnant. Then it flew up to the eaves of the nearby garage. Here it entered, disturbing the House Sparrows who possessed this roost. A few minutes later, H-35 carried some feathers to cholla 22. Nothing further was done on these tentative starts until 26 December, when work on nest 23B was resumed. The wren completed this nest and occupied it. Until it made the final decision, we believe the wren roosted under the garage eaves. All the chollas mentioned were in the center third of our lot. Their sites formed an isosceles triangle 60 x 120 x 120 feet (18.3 x 36.6 x 36.6 m). Here, apparently, was a firm choice of area, but there was considerable uncertainty in regard to the particular site (Fig. 3.4).

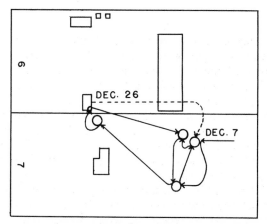

Fig. 3.4.   Movements of Cactus Wren in search of roosting nest site on 7 December, 1941, in Lot 7 on Kleindale Road, Tucson, Arizona. Final choice made on 26 December.

The actual start of a roosting nest may take place at any time of day. For example, on 4 October 1941, a wren began to lay the foundation of nest 1D at 1500. Sometimes an unpleasant disturbance at a roosting nest after dark, such as being trapped, may cause the wren to seek a new site the next day. Once the nest has been started, however, construction work usually begins early each morning, sometimes before sunrise. The first 2 or 3 hours are the busiest. Then work slows down, seldom continuing until noon. There is occasionally another period of activity, usually a brief one, in late afternoon. It would be logical to suppose that the rate of construction would vary with the urgency; perhaps it also varies with the individual. There is some evidence in our incomplete records which suggests that cold weather may be a factor influencing the rate. Table 3.2 shows the date of beginning construction and date of occupancy of ten roosting nests for which we have sufficient data.

Table 3.2. Beginning of construction and date of occupancy of roosting nests. Kleindale Road area.

Nest	Date begun	Occupied
76A	3 July	7 July
1F	15 July	20 July
92C	22 July	23 July
4C	5 August	7 August
82A	10 August	10 August
1D	4 October	6 October
14B	18 October	19 October
73F	8 November	9 November
1E	1 December	1 December
1A	19 January	20 January

The average date of occupancy is 2.7 days after the start of construction. Nest 82A, begun on 10 August, and occupied that same evening, was built upon a foundation which had been laid on 29 July and then abandoned. It must not be inferred that these nests were completed at their time of occupation. Occupation usually began as soon as the nest cavity had been rounded and outlined somewhat overhead. It was hardly more than a shell at this time, with a lattice of grasses for a roof. The outside finished appearance was attained at the end of 7 to 10 days. The lining of the interior took longer and might continue in irregular bursts of activity for several weeks.

Nest 25B proved to be a disconcerting exception. The wren began it on 31 August 1952 with a few straws. The next day work proceeded so rapidly that by 0918, when the wren stopped, the nest, although rather thin, was entirely covered over. The wren slept in it that evening. In this instance, the rapid rate of construction could hardly be attributed to desire for a warm roosting place. The afternoon maximums of temperature during this week reached 108F (42.2C), and the nights were far from cool. Neither does it seem probable that the wren hurried to finish work in mid-forenoon because it knew the day would be hot.

Attentiveness in the construction of the roosting nest was extremely variable. Distractions occurred frequently, stopping work for considerable periods. These distractions or interruptions did not always come from human disturbance in the vicinity. Noisy automobiles or children, of course, had their effect upon the rate of nest construction. Other wrens were often the cause of work stoppages. It was not unusual for the builder to leave his unfinished nest to join the small territorial group of wrens roaming the area in search of food. Perhaps it was time for a "coffee break" anyway, but it could have been a natural, aggressive tendency to seek a share of the food supply located by the neighboring wrens.

Nest 23A, watched on 17 August 1941, revealed the following active periods, beginning at 0720 (construction on this nest doubtless had started earlier than this). The wren made 18 visits to the nest in 14.5 minutes; then it was absent 5 minutes, after which it made 21 visits to the nest in 18.5 minutes. Our observations were interrupted at this point for nearly 40 minutes. Then beginning at 0838 the wren made nine visits to the nest in 10 minutes, stopping at 0848. No further building was done until 1705, when the bird returned and resumed work, finishing at about 1730. We found this nest occupied in the evening.

We watched nest 25B from 0836 to 0918 on 1 September. H-56 tapered off its morning work as follows: it made six visits to the nest in 8 minutes; then it was absent for 14 minutes. After this it made five visits to the nest in 5 minutes; then it was absent 12 minutes, after which it made three visits to the nest in 3 minutes. Work then ceased for the day.

Construction of nests began with the placing of material at the far, inner end of the space chosen among the cholla joints. When available, the wrens used dried, stiff stems of buckwheat (*Eriogonum trichopes*) at first; then came small tufted grasses such as *Tridens pulchellus* and *Schismus barbatus*. Both of these are abundant in the area. Buckwheat formed a framework to which smaller bits of material could easily be secured. The first part of the floor went down upon the spiny joints; this was followed by a gradual filling in of the back, until a slipper-like toe evolved. Now and then the sharp, dense spines would apparently get in the way, for the wren pulled and yanked in an attempt to break them. At no time did we observe the wrens breaking off spines preliminary to the actual installation of nest material. Eventually the floor reached a thickness sufficient to cover the spines. Their removal then would thus appear unnecessary.

By the time the roof of the nest cavity was outlined, the location of the entrance became evident. At first, material might be brought in from the top or sides or front. Once, after placing some grasses on the floor, a wren left the nest by climbing through the newly installed roof.

Soon, however, the nest took the shape of a wide-mouthed jar, sloping down inwardly and facing outward from the cholla. Short grass stems, 2 to 3 inches (5 to 7.6 cm) in length, were poked directly into place with the bill, after which the wren turned and pushed with its body in several directions, packing the material in and at the same time widening the nest cavity. Sometimes it turned completely around and used its feet to scratch the loose material farther back. The end of a longer grass stem would be trampled and anchored to the floor; then the remainder would be fashioned into the wall and roof, the wren standing upright, reaching to the roof and poking the stem into the required curvature.

As work proceeded, the nest grew in shape and size from the inside. The cavity expanded, and its walls became denser and thicker with the addition of shorter bits of grass. The lining normally consisted of finer grasses, small feathers, and some plant down; the latter, in this locality, was chiefly the pappus from the achene of the desert broom which matures in late autumn. The vestibule, originally about 3 to 4 inches (7.7 to 10.1 cm) in its outline dimensions, was filled from the inside and lengthened to a neat cylinder of 1.5 inch (3.8 cm) bore, with walls perhaps an inch (2.5 cm) thick. This tube tapered rather abruptly to its entrance. The outside of the nest came to appear more and more as if the larger grasses had been wound around it. Only at the entrance did the grasses protrude unevenly. Frequently, it was the panicle end which had been pulled in, leaving the larger, stiff end exposed. As the vestibule lengthened, it might drop to a horizontal position or even slope downward or sideways as it followed the direction of the doorstep joint of cholla. It was not unusual to find the vestibule almost incased in the spiny cholla joints. No wren could enter a nest without leaning forward and practically crawling through the tube. Even so, its back must be scratched at times. The weights of two representative roosting nests were 4.7 and 4.8 ounces (133.5 and 136.5 g).

The foregoing applies to the construction of nests before the Kleindale Road neighborhood grew into a crowded residential section. Nests in later years were built of any available materials in the vicinity. These new materials were seldom as satisfactory as the old. Nests became larger, more ragged in appearance, and they lacked the firm, woven texture of the earlier ones. Heavy rains might collapse them, and thrashers had little difficulty in tearing them apart. In general, their useful life was shorter. The lining, now entirely of chicken feathers, often virtually filled the nest cavity. Indeed, some nests were merely thin pouches stuffed with feathers.

It was seldom necessary for the wrens to search far for nest material. Frequently such material was available immediately below and within a radius of 15 to 25 feet (4.5 to 7.5 m) of the nest. In spite of the abundance of man-provided trash, there was evidently a strong compulsion to build first a framework of long grasses and weed stems. Wiry Bermuda grass (*Cynodon dactylon*) runners and the thin, long, much-branched culms of mesquite grass were pulled and jerked until broken off; then these were carried singly to the nest. Shorter pieces might be close by, but the longer ones were selected first, even at the expense of considerable effort. Often the effort was so vigorous that when the stem snapped, the wren tumbled backward, unable to balance itself with its spread wings and tail. Smaller bits of material were gathered up from the ground until the bill would hold no more. As construction proceeded the quest for material broadened. We have seen wrens carrying lining material as far as 200 feet (61 m).

Cactus Wrens were not averse to taking material from old damaged nests, but the parts used were chiefly scraps of the lining. Apparently grasses for the exterior could be obtained more readily from the ground. Old House Finch (*Carpodacus mexicanus*) nests sometimes provided cotton-like shreds for lining. We have no records of any Cactus Wren tearing down one of its own nests, either one just begun or an old one, in order to use the same material to construct another nest in a different location. The pieces of a nest, abandoned in its initial stages of construction, were never transported to another site.

During the first few days of construction, the actual time spent in the nest arranging the material was very brief. The average time for 18 visits to nest 23A on 17 August 1941 was 7.6 seconds. The minimum time was 4 seconds, the maximum 15 seconds. Another wren, working harder and longer, on nest 25B on 1 September 1952 devoted from 15 to 30 seconds to inside work. This latter nest was noticeably farther advanced in construction at the end of the first 2 days. The lining requires more time to arrange. Perhaps the urgency is not as great, once this stage has been attained, for the nest is now in use at night. A minute, or even occasionally as much as 2 minutes, were recorded before the wren came out. Perhaps the builder was simply resting comfortably inside while the lining settled into its proper place.

Various observers have tried to determine what external factors, if any, influence the direction of the nest entrance. Bailey (1922: 167-168), working at the base of the Santa Rita Mountains, arrived at no positive conclusion (Table 3.3). In Table 3.4 we show the directions faced by 122 roosting nests on Kleindale Road, 182 nests on the Santa Rita Experimental Range, and 508 nests in Saguaro National Monument. Evidently there is no preference as to the direction in which the nest is faced. Ricklefs and Hainsworth (1969: 35), using a sample of only 88 roosting nests, have suggested that "The random orientation among roosting nests may be related to the low wind speeds at night." We have no proof of this, and can see no way of demonstrating that wind speed is considered in the construction of a roosting nest.

Cactus Wrens are peripheral nesters. The nest, when placed in a dense jumping cholla, is almost always at the outer surface of the crown, the entrance pointing outward from the cholla. A wren does not climb through the branches to get to its nest. If it approaches from a direction opposite to the entrance, it will invariably fly in a circle and enter from the outside, or it will land on the top of the cholla and fly down to the

Table 3.3. Orientation of Cactus Wren roosting nest entrances.
(Near Santa Rita Mountains, summarized from Bailey, 1922.)

Direction	Nests on 53 Acres	Nests outside area	Nests in cholla cacti	Total
N	4	4	2	10
NE	5	3		8
E	1	2	4	7
SE	3	6	4	13
S	3	4	8	15
SW	12	5	8	25
W	2	2	1	5
NW	7	1	4	12
TOTAL	37	27	31	95

Table 3.4. Orientation of Cactus Wren roosting nest entrances.

Direction	Kleindale Road	Santa Rita Experimental Range	Saguaro National Monument	Total
N	8	33	72	113
NE	12	17	41	70
E	30	21	80	131
SE	9	16	57	82
S	27	30	85	142
SW	13	20	56	89
W	14	25	62	101
NW	9	20	55	84
Total	122	182	508	812

nest. There are good reasons for this behavior. First, it would be difficult and discouraging to drag the sometimes long and intricately branched grasses and weed stems through the maze of spiny twigs and branches to a nest facing the interior. Second, and this may be even more important, the outward-facing entrance provides a greater field of view and a quick means of escape in the event of approaching danger. Cactus Wrens experience no trouble in climbing about in the spiniest of chollas, but they move slowly. Were it necessary to dash suddenly out of a nest and then through several feet of cholla twigs to reach flight freedom, the probability of impalement would be great. In the less spiny cane cholla, where the side branches are fewer, or in the older jumping chollas, whose lower joints have been lost, nests are frequently placed close to the trunk or at the ends of the stems. In such cases, any horizontal entrance would then face outward and away, for there are no obstructions in front. Once a nest site has been chosen, convenience, accessibility, and safety dictate the position of the entrance. The best direction is outward (Fig. 3.2).

4. Vocalizations

Song. Many early observers had difficulty in describing the Cactus Wren's song. Heermann (1853: 263) began, by reporting the wren's "uttering at intervals a loud ringing note." Others, apparently having no firsthand information of their own, repeated this statement in later publications, and then made various confusing and contradictory additions of their own. More recently, Woods (in Bent, 1948: 229-230) said "the voice of the Cactus Wren has rather a deep, throaty quality, sometimes becoming almost a croak. The bird uses a great diversity of notes, some of them grating or ratchetlike, varied with jay-like squawks and occasional cries suggesting the plaintive demands of young birds. While foraging, a softer clucking or croaking note may be given at intervals . . ." Song "is the rapid repetition of a single staccato note. The quality of this note varies, but never in the same series . . . The most tuneful utterance that I have ever heard from these Cactus Wrens was a warbling song given by an immature bird, a song so soft that it could have been heard only within a distance of a few feet." Finally Brandt (1951: 184) approaches a more complete and accurate account as follows: "incessant, mechanical singing . . . a succession of sharp, staccato notes, as though he were scolding . . . 'riv-riv-riv-riv' notes, always in the selfsame key and so rapidly one could not count them audibly. The series ran from 8 to 12 notes with a considerable pause between each group." Another wren sang "from 12 to 18 notes in each run . . . Close by there is considerable roughness and harshness in its voice, which, however, becomes mellowed by distance and loses its mechanical rasping . . . In addition . . . [there is] a series of coarse, scolding notes similar to those of the House and the Long-billed Marsh wrens, which is entirely unlike the territory song."

Thus it was nearly a hundred years after the report by Heermann of the discovery of the Cactus Wren near Guaymas, Sonora, Mexico, that any accurate description of its song was published. By song, we mean the vigorous vocal expression that coincides with the establishment and maintenance of a Cactus Wren's territory. It is never musical in the ordinary sense, for it is harsh and frequently grating. It is a series of staccato syllables, the first three or four uttered in a low tone, the next few reaching a steady, greater amplitude. This level is then held to the abrupt stop at the end, the entire song lasting only about 4 seconds. After a pause of from 4 to 8 seconds, the series of ten to 12 syllables is repeated, and so on, again and again. It is difficult to assign an initial consonant to each syllable, if indeed there be one, but the vowel can be *ä,* short *a, i,* or *u.* The *r* sound is very strong, even suggesting the grinding of pebbles among one's teeth! Perhaps the best wording would be *char-char-char-char-char* or *rar-rar-rar-rar-rar,* the vowel varying in different songs and birds to *ä, i,* or *u.* Occasionally a short 3- or 4-syllable song is heard.

In singing, the wren elevates its bill only slightly above the horizontal. Most of the drawings in the literature depict the wren with its bill far too close to the vertical. The song is produced with such vigor that the feathers of the throat stand out and quiver from the internal vibration. It is loud and penetrating, and is easily heard at a distance of 1,000 feet (305 m). As Brandt reported, distance removes some of the harshness, giving it at times a sort of ringing quality.

Buzz. — This danger note is a true buzz of varying intensity. It doubtless could be duplicated easily by means of an electrical buzzer. It may be low and persistent while the wren is following a cat or a Roadrunner (*Geococcyx californianus*). It can become a frantic, louder, half-screech if one approaches the nest and handles a frightened nestling.

Tek. — This warning call is a rapidly uttered staccato series: *é-é-é-é-é-é.* The vowel may change to short *i* or *u.* It is usually given when the nest or fledglings are in mild danger, being succeeded by the *buzz* note as the threat increases. At first, the syllables may be far apart, sounding like *check* or *tek;* then they lose the consonant sound as the series gains momentum. Sometimes the call suggests the *put-put* of a gasoline motor-driven pump in the distance! Outside of the nesting season, it may serve partly as a location call, although sometimes we suspect it may be used as a warning to intruding wrens.

Rack. — This call seems to be for locating the position of the mate or other wrens. It is sometimes the first note uttered when a wren comes out of its roosting nest in the early morning. The initial *r* sound is very strong here, and it is rather prolonged, suggesting the beginning of a growl. The vowel sound is often varied to long *e* instead of short *a,* producing *rrreek* instead of *rrrack.* Two or three calls, deliberately uttered, with a short pause between, are all that one usually hears at a time. However, sometimes, as though in excitement, a series of six or more may be given, the last part with increasing tempo. Occasionally the note is modified to a sharp *tirrip* or *turrup.*

Scri. — This peculiar, scratchy note occurs chiefly during territorial boundary disputes. It is very rough and harsh, almost impossible to describe except by comparing it with the sound produced when a metal-strip rake is dragged rapidly through loose gravel. Several of these calls are usually uttered as the dispute begins. They also occur during the fight and pursuit.

Growl. — This note is part of the recognition display. It sounds exactly like the thing the name defines — a distinct *growl,* in which the *r* sound predominates. It may be described as *rrrraawrr.*

Squeal. — The purpose of this seldom-heard sound is obscure. It is uttered by the male in the vicinity of his nest, before a breeding nest has been started or completed. The female is always near by. The sound has a very painful quality, but there is no evidence of any physical injury.

Peep. — Nestlings emit faint *peep* notes.

Dzip. — Fledglings have a quickly uttered *dzip, dzep,* or *dzup* given singly at short intervals.

The fledglings' subsong, rehearsed song, and acquisition of adult warning and danger call notes are described in Chapter 14.

Wolford (1969) has recorded the vocalizations of the Cactus Wren in the Tucson region. The sonograms reveal the important physical characters of a repertoire entirely different from that of other species of wrens in southern Arizona. We hope that a similar study can be made, for comparison, of the eleven closely related southern members of the genus *Campylorhynchus* in Mexico, Central America, and South America.

5. Pair Formation and Establishment of Territory

Unfortunately we do not have any exact data on pair formation, for we never witnessed the first meeting of a male and female Cactus Wren. Whenever an adult disappeared, another took its place. The introductory ceremonies must be very brief; we were seldom aware of the substitution until it was complete. Nevertheless, from later observations, we believe it is possible to offer a probable explanation of what actually takes place at this important time. To the human eye, the sexes of the Cactus Wren are identical in coloration and size. We doubt if any external character is a factor in pair formation. Behavior, then, must be the key to sex discrimination and subsequent pair formation. This view is not original; it has in the past been applied to other species of birds. The element of recognition is involved in four possible situations.

1. Male meets male. The loud, persistent song of the male, announcing his ownership of the territory, serves as a warning to other males. They respond by similar songs from their own staked-out territories. These songs can be considered as part of an aggressive, hostile act, somewhat similar, at least in effect, to the barking of a dog as it threatens the coming postman. Singing usually increases as two males approach their common boundary. If the conflict erupts into active chasing or physical combat, the singing may temporarily stop, to be replaced by the scratchy call note, but, as the rivals retire, it is again resumed. When male meets male in the spring, we can assume that they will sing, then quarrel. In human analogy, of course, the singing is vocal intimidation or warning. Whatever the intent of the song, it would seem illogical to suppose that the effect is solely hostile. The song is admirably suited to the recognition of sex, for it is a distinctive form of male Cactus Wren behavior.

2. Female meets male. In this case the song alone would be sufficient for sex determination on the part of the female, for the female does not regularly sing. When at times she does sing, usually immediately before copulation, her song is weaker and often at a slightly higher pitch. Observa-

tion readily reveals that the female is attracted to the vicinity of a singing male, and frequently flies to him.

3. Male meets female. This is a more complex situation. If a female trespasses, we can assume there will be no song to identify her. It would seem natural for the male to fly toward the intruder to settle the matter of identification at once. What occurs then is probably what we have recorded time after time when a male flies to a female already paired to him. We have designated it as "display-growl." As the male arrives, for instance, on a fence post on which his female is perching, he spreads his wings and tail in a threatening gesture, uttering meanwhile a growling sound. This sound may be of one or several short syllables. The female also displays at the same time in a similar manner; then she usually crouches. She possibly growls also, but the display is so rapid that the sounds often seem simultaneous. The duration of the display is only 2 or 3 seconds. The female is the first to return her wings to normal position. The male retains his threatening posture a moment longer. He may then poke his head under her chin, or peck under her tail. Sometimes he pecks her lightly on the head or rump. Occasionally, as she crouches, she pecks his toes. In all cases observed, the attitude of the male was one suggesting dominance. The female cringed. As the display ended, the female usually flew down and searched for food on the ground. The male remained a while on his perch as though on guard. The most obvious conclusion is that the display is a threat. The response of the other wren determines whether there is a conflict, or friendly submission. In other words, the response reveals the sex. These displays occur throughout the year. They seem to be a necessary form of greeting between the male and female of a pair during the period of active territorial ownership. They reveal the probability that the male is unable to recognize his own mate except by testing her each time they meet. This may be an extreme view to take in regard to recognition. An alternative is to suppose that continued association with the one female gradually softens the challenge of the display to a ritual, assisting in pair-bond reinforcement. The challenge may become a greeting under the continued, stereotyped submission of the female.

4. Female meets female. The female probably recognizes another female, not as one of her own sex, but as one which does not attract her. There is no song to follow or display to which to respond. The effect is neutral, until a conflict develops in relation to a male. Then the more aggressive female drives the other female out of her territory.

It seems safe to conclude that the pairing bond is accomplished rapidly and without elaborate, lengthy ceremony. Briefly, a female Cactus Wren is attracted to a singing male. He immediately threatens her. Her submissive response is all that is required for pair formation. From then on they remain together.

The above is an ideal, simplified explanation, drawn by inference from our observations of the behavior of wrens after they are paired.

We have, at least, been certain of the sex of many of our local wrens. The behavior of color-banded individuals has been studied at length. The question of which is male and which is female is easily answered at the time of copulation, for the wrens at this time are typically avian in their behavior.

The four possible situations, which we have outlined, apply only to the meeting of two individuals. Where several wrens are present, variations are to be expected. If all the females were at the same stage of the breeding cycle and were equally "attracted," it could be assumed that all would fly to a singing male and take part in the display. In our area this did not always occur. During the prolonged period of attempts at eviction of HF-35 by HM-23 and HF-30, in the winter of 1942, we did not observe any displays by HF-35 and HM-23, nor did we ever see the female, HF-35, fly to the male when he sang. True, we could have missed such events, for our observations were often interrupted and discontinuous. Yet the behavior of HF-35 suggested that she was usually in a cowed, intimidated state, unable to attempt courtship. She was not equal to HF-30. She could be considered third in the peck order.

Where females were equally attracted toward a male, their behavior was sometimes more difficult to follow because of the multiple action. In the several cases observed, we can assume that the process of pair formation was at least going on, if it had not already been accomplished. On 21 December 1952, H-62 sang a number of times from the top of an electric power pole at our east fence. HF-57, who had perched on a wire a few feet below H-62, tried twice to fly up to the other, but she was gently pecked each time, until she dropped back and clung to the side of the pole. She finally flew to the ground. Fifteen minutes later, HF-58 tried the same approach, but she, too, was pecked. H-62 remained with us only a short time. Its sex was not known, but it behaved like a male. In the above case, H-62 was apparently not ready to choose a mate. The absence of a display in this instance makes this a troublesome deviation from the "ideal" situation pictured when male meets female. Perhaps it can be best explained by the fact that the male had both females in his field of view at the time. There was no sudden appearance to trigger a threatening posture. Furthermore, at this time, the group of wrens in the area had not yet chosen definite territories.

Later, on 25 December 1952, a more extended affair occurred about 1030 in lot 4. Four wrens were in a cholla. Although the entire action could be observed, the birds were unfortunately too far away for positive individual identification. Three in a lower part of the cholla were attempting to climb up to the fourth, which perched on the top joint. One of the lower ones, carrying some grasses, tried twice to reach the top. The display-growl could be heard several times. Then the top wren flew westward to another cholla, followed by the others. The leader uttered a *tek* sound frequently and twitched its wings and tail. Meanwhile, the other wrens moved upward, causing more displays and growling sounds.

They never actually reached the top, for the cholla joint afforded room for only one bird. The displays and movements were rapid and difficult to follow, and they were even more difficult to note down in order. The group moved from cholla to cholla, westward about 200 feet (61 m), stopping at least eight times, and repeated the same behavior in each cholla. At the last cholla, the leader sang. Immediately the other three wrens climbed up, causing more displays. Only one explanation seems possible here. Three females, equally advanced in sexual development, were attracted to this one male. Evidently territorial intolerance was slight, for only a single brief chase was observed in one of the chollas during the entire action.

On 3 January 1953, another singing wren, a noband, attracted two females to the same electric pole at our east boundary. Again, twitching of wings and tail and general fidgeting occurred, and again one of the wrens was pecked until it left. On 14 February 1953, HF-58 flew to a cholla in which HF-57 was climbing about. At once the latter wren spread its tail as though threatening, but there was no further evidence of warning. They both began searching for food in separate directions.

The season was slow in 1953. The first breeding nest was not begun until 1 March, and, at this time, wrens were still undecided about their territories. Females apparently outnumbered males. Singing by females was more frequent than in other years. It often confused the work of identification. Evidence of the male's reluctance to drive out other females, or his inability to recognize his own, was provided on the evening of 6 March. HF-57 and her noband mate displayed and growled from cholla 5 in the north part of lot 7. Five minutes later, the male sang from a mesquite tree in lot 8. Then HF-57 sang several times from a post in lot 6, after which she retired in her roosting nest, 21B, just north of our house. Soon HF-58, an unattached female, began singing from lot 6. Noband male appeared at the unfinished breeding nest 25C, located 85 feet (25.8 m) north of nest 21B, and uttered a peculiar squealing or whining sound. Then HF-57 arrived and the display-growl occurred. Both then flew to the pole on which HF-58 had been singing, causing the latter to move on to the electric wire. Noband male *tek*ed, while HF-57 twitched; HF-58 moved 3 feet away (.9 m). HF-57 moved closer, but HF-58 remained, although she had stopped singing. HF-57 flew back to her mate. There was a brief *growl* sound; then noband continued *tek*ing, with HF-57 still fidgeting. Suddenly HF-58 flew north to the ground, followed quickly by the other two. There was chasing on the ground among the creosote bushes about 50 feet (15 m) north of the fence; then scratchy sounds were heard. HF-57 returned to the pole and sang a number of times. She then flew northwest about 200 feet (61 m) to another pole and sang repeatedly. Her noband mate flew to the first pole and also sang.

Evidently at this stage the male was able to keep track of and maintain recognition of his mate, for he did not fly to her. Soon HF-57 flew back to her nest 21B and retired again. Meanwhile, noband flew northeast to a cholla where another wren had sung briefly. Here again was the display-growl, but the new female could not be identified. It may have been HF-58, for this wren now appeared in front of nest 25C. She made some low sounds and seemed disturbed as she peered into the entrance. At once, HF-57 arrived. There was a sharp *growl* as both birds dropped out of sight behind the cholla. Soon afterward, HF-57 sang again from the pole in lot 6; then she retired in nest 21B. By this time it was quite dusky, and the other wrens had become quiet.

To summarize, noband male displayed with two females; HF-57 left her roosting nest twice in the evening, once to drive HF-58 away from the territory, and once to protect her breeding nest from HF-58. No further data on this triangle could be secured, for both HF-57 and HF-58 disappeared in the course of the next few days. The noband male, or another one, paired with HF-59 soon thereafter; both completed nest 25C for their breeding nest.

The wing and tail twitching, accompanied by the rather slowly uttered *tek* note, apparently did not occur often in the early years of this study. We suspect it is more prevalent when there is a surplus of females. Both sexes take part. The male reacted in this manner when confronted by two or more females. The female reacted similarly when she was faced with competition. It is probably a form of threat behavior, although to the human eye it suggests nervousness.

Territorial assertion in the Kleindale Road block, if not present earlier, must begin at least in January, for it was in this month that wrens other than the pair gradually disappeared. By 15 February, a pair was usually in complete possession of its breeding area. Both sexes assisted in this clearing-out process. The pair which had frequented our lot and the adjoining one to the west in the course of the preceding year seemed to dominate the situation. Others might crowd in in late fall, but eventually they would be forced to vacate. Obvious fighting was rarely evident. Rather it was the persistent, nagging movement toward other wrens, sometimes leading to active chasing, which brought about the expulsion of the undesirables. These threatening runs and chases were seen most frequently in January, 1942, when six wrens were present in the area.

At the beginning of the year, HM-23 and HF-30, already in possession of the territory, appeared to dominate it. HM-37 and HF-38 moved out in the first week and settled nearby in the northeast part of the block just outside the territory of the dominant pair. H-36, endeavoring to remain, carried nesting material to a hole under the eaves of a neighbor's garage. The resident female, HF-30, chased it frequently. These chases usually occurred when H-36 approached the nest hole with some chicken feathers, and they seldom extended more than 10 to 30 feet (3 to 9 m). Afterward we saw no antagonism as the wrens foraged on the ground. Nevertheless, the effect of the interference seemed to be cumulative. If other wrens were in sight, H-36 appeared more and more fidgety and nervous as it approached its roost. Once HM-23 dashed toward it as it landed on the garage roof, driving it off, but generally his mate was the more aggressive.

On 4 January, a brief fight occurred on the ground near the garage, but the action was so swift and confused that it was not possible to trace its cause or course. When the two wrens separated to a distance of 20 feet (6 m), HF-30 held a piece of cotton or a feather in her bill. Two days later, we saw HF-30 clinging to the garage wall, attempting to pull out some of the projecting nest material from the eaves. Meanwhile, H-36 buzzed in protest from a safe point about 50 feet (15 m) away but did not offer to defend its roosting place. On the 16th, we found it lying dead on the ground just beneath the nest opening in the garage. Whether it had been attacked and killed by the other wrens, or by the House Sparrows, which also roosted under the eaves, or died from natural causes, we were unable to determine.

The remaining outsider, HF-35, held out until the middle of February. On 3 January, we saw HM-23 following her about in the yard, HF-35 always keeping some distance ahead. She never fought back, yet she would not leave. She continued to work on her roosting nest right in the midst of the others' territory. Once, when HF-35 was at, but outside, her nest, HF-30 landed in the entrance, facing into it, wings and tail spread, as though intending to block ingress. In a few moments she moved to the top of the nest, stretched her neck upward, her pose suggesting the upright, alert stance of a Round-tailed Ground Squirrel. HF-35 waited quietly on the ground below. When the other finally departed, she resumed work on her nest. She showed a remarkable persistence and tenacity in clinging to her bit of ground at great odds. Not only was she chased and disturbed by both HM-23 and HF-30, but her roosting nest, nearly completed, was torn apart by a Curve-billed Thrasher. She began another nest. This, too, in a few days was damaged. She began work on a third, left it, and started a fourth nest. On the 22nd another confused fight, with squeals of pain, took place in the front yard under a creosote bush. Although the participants scattered quickly, so that again the action was left in uncertainty, we feel sure that HF-35 had been attacked by one or both of the resident birds. Chasing continued for the next 2 weeks. After that HF-35 moved northeast to the catclaw growth along the Rillito, beyond the area occupied by HM-37 and HF-38. Here she found a mate. Again an element of uncertainty creeps in, for her departure could have been induced by the discovery of the unattached male near the Rillito, and not by the persecution of her neighbors. She had shown abundant reserves of strong passive resistance to eviction.

A puzzling aspect of the late winter chasing was the apparent disinclination to follow it through to the territorial boundary. As previously mentioned, the chases were almost always short, seldom over 25 or 30 feet (7.5 to 9 m), and they often stopped quickly. A threatening run toward another wren might be as little as 10 feet (3 m) in length. When the threat ended by the stopping of the chase, the second bird also stopped. Usually, thereafter, both birds continued foraging on the ground without apparent hostility. Sometimes it was hours later before another such event took place. These threatening gestures of January 1942 occurred almost in the center of the territory of HM-23 and HF-30. The mere presence of an outsider did not always provoke a dispute. However, the construction of a roosting nest seemed to be considered more of an intrusion, and it brought forth a greater and more vigorous reaction.

The following winter, when HF-35 had again moved into the local territory, we saw another chase at our east fence. HF-35 came through the fence, followed by HF-39, another female, now the new mate of HM-23. HM-23 perched in the top of a nearby cholla and did not take part in the chase. HF-39 suddenly ran toward HF-35. The latter moved about 3 feet (1 m) away, and was again chased. This time she flew to the top of a fence post. HF-39 flew up to the next post, 10 feet (3 m) away, and then flew directly at HF-35, forcing her to leave the post and drop to the ground. From this point she was again chased, through the fence, back in the direction from which she had come. The entire distance covered by the two wrens did not exceed 50 feet (15 m).

If any intruding male ever ventured into the territory, it must have been along the remote perimeter where we failed to observe it. We never saw a foreign male take up a singing position within the territory. The reaction which might result from such a situation remains unknown to us. Boundary disputes were frequent during the crowded years, but actual invasions were not observed. Thus the establishment of ownership consisted of evicting the females which naturally were attracted to the dominant, singing male. This work fell mostly to the resident female. The male was not always cooperative in disposing of an additional female in his territory.

The hostility, so evident toward members of its own species, did not extend often to any of the other birds which gathered into loose winter flocks around our home. There might be an occasional quarrel at the feeding table, but, as a rule, each species waited its turn according to size. The resident House Finches, House Sparrows, and Black-throated Sparrows (*Amphispiza bilineata*) gave way to the larger Cactus Wrens. If a Curve-billed Thrasher or a Gila Woodpecker arrived, the wrens edged to the side, and then returned, when they were permitted, to eat what was left. Even when nesting activities began, the tolerance of other species continued. Numbers of wintering White-crowned Sparrows, Brewer Sparrows (*Spizella breweri*), and Lark Buntings (*Calamospiza melanocorys*), which frequented the territory from October to late April, were ignored. In early spring, Brewer Sparrows, House Finches, and even Curve-billed Thrashers sometimes roosted in the chollas which contained occupied wrens' nests.

Coincident with the harassment and eviction of all other wrens, except his mate, was the increasing frequency of the territorial song of the male. Singing, to be sure, occurred to some extent on practically all days of the year. It was lowest in frequency during November and the first half of December. A fresh winter rain, or above normal temperatures in the latter part of December, produced a noticeable increase at once in song. By January, it was evident that ownership of the territory was being advertised. We were never fortunate enough to observe the first arrival of a male on a vacant territory, and his immediate endeavor to establish residence and obtain a mate. In our area, at least one pair was always present during every winter of this study. Since the male already had a mate, his singing must necessarily have been almost exclusively for the purpose of proclaiming ownership of his land. As such, it did not always have the fervor and intensity of the song which might characterize his search for the replacement of a lost mate. Apparently the song is stronger when a female has not yet been secured.

Singing began before sunrise, even on cold, frosty mornings. It was not unusual to hear a few songs during an early morning rain. Sometimes singing started spontaneously, but frequently was in answer to the song of another wren nearby. Seldom did the songs overlap. After each song there was a pause of sufficient length for the wren to listen to the other's song before beginning another of his own. It could be called antiphonal singing, but it was performed by two rival males, not by the male and female of a pair. Of course, if a third singing male was within earshot, overlaps became inevitable.

At awakening, singing always preceded the search for food. After a preliminary *rack* call note, a series of songs was heard from perches close to the nest. Once, to our surprise, as we watched a nest, the male uttered his first song from inside the nest just before coming out. Early in the season, songs were few in number. Later, 25 to 30 or more songs might be sung before feeding began. The male sang from most of the available perches in his territory, such as creosote bushes, mesquites, roof-tops, radio antennas, windmills, and electric poles and wires.

The singing stations were most numerous in lots 6 and 7, where the roosting nests were usually located. They decreased noticeably along the perimeter of the territory. At the far boundaries we observed no singing at all, although conspicuous elevated perches were present. In other words, the maximum territory extended somewhat beyond the singing stations and the boundaries were not advertised by means of song. Infrequent singing occurred to the north and east. This area, reaching as far as Greenlee Street and to the edges of lots 1 and 10, was, nevertheless, utilized for food and nest materials. South of Kleindale Road, the 10 acres of uniform creosote bush were occasionally visited by the wrens, but they were seldom advertised as acquired territory. As previously mentioned under roosting nests, the center of activity was confined principally to the southwest corner of the tract. There seemed to be a definite reluctance to frequent the area west of Flanwill Street where chollas were few in number. As shown in Fig. 5.1, the singing stations during January and February of 1944 were grouped chiefly in a semi-circle east of cholla 6.

Defense of territory. Up to 1947 the territorial boundaries were so far distant from our house in lot 7 that we never saw or suspected that boundaries were actually defended after the territory was occupied. It appeared to us that no defense was necessary, once the breeding season began. Other territories were far removed; the breeding nest of the Edith Street pair in territory II was at least 600 feet (182.9 m) from the nest

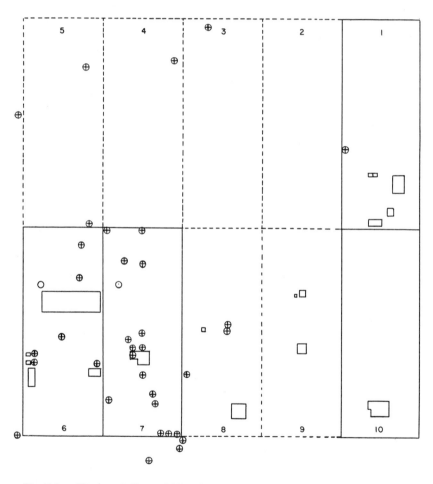

Fig. 5.1.   Singing stations of HM-23 in January and February of 1944 in the Kleindale Road area, Tucson, Arizona. Circles indicate roosting nests; crossed circles show singing stations.

in territory I. The area utilized by the local pair was probably the maximum it needed. We occasionally heard the scratchy *scri* note connected with disputes. Adults sometimes used it in evening squabbles at roosting nests, and also in early spring pre-nesting activities. The division of the 10-acre (4 ha) block into two territories in 1942 and 1945 was not a drastic alteration. The newcomers in the northern part of the area obtained the less desirable portions of the tract. The breeding nest of territory III in 1942 was approximately 310 feet (94.5 m) from that in territory I; in 1945 it was 435 feet (132.6 m) away.

Defense of territory is a somewhat misleading, elastic concept. Defense occurs, of course, but in our studies it was always accompanied by a compromise. Each of three annual (1942, 1945 and 1947) intrusions brought about the annexation of a portion of the original territory. What would have happened had a third or a fourth pair attempted to move in is uncertain. There must be a limit beyond which a Cactus Wren refuses to give up any more land. This limit may have been reached in 1947, when the breeding nests of territories I and III were only 180 feet (54.9 m) apart. Boundary disputes became very frequent; they began in the spring, and they continued through the entire year and even into the following spring. It seems incredible that we could have missed these conspicuous, noisy disputes before. We can only assume that they occurred infrequently because of the greater separation of the breeding nests.

In the early part of 1947, we observed nothing indicating that a change in the local distribution of wrens was in progress. HM-48 and HF-49, whose roosting nests were in the rear third of lot 7, seemed to be in complete possession of the 10-acre tract. Then, on 19 February, we suddenly discovered that another pair of Cactus Wrens had occupied the southern portions of lots 6, 7, and 8. (We trapped and banded them later, and numbered them HM-54 and HF-50.) They were building a new nest, 19C, in a cholla 12 feet (3.6 m) southwest of our front porch. The female, at least, of this pair probably came from territory II, for she returned to that area nightly, until she had built a new roosting nest, 8P, in lot 8. If any quarrel or fight occurred over the annexation of this considerable portion of the original territory, we failed to observe it. As often happened, when we became aware of a change in the situation, it was already an accomplished fact. It should be noted that in this instance there were no roosting nests in the land that was given up to the invaders. Had there been roosting nests, the outcome might have been different.

After his retreat northward, we heard singing more frequently from HM-48. The new male, HM-54, busy at nest building, was not at first as vigorous a singer as HM-48. Most of his singing was done while at work, at or in the vicinity of his nest. Later he found time to give rebuttal to HM-48, chiefly from stations in the trees in the southern portion of lot 7, and from electric wires along Kleindale Road. On 22 March we observed the first indication of a definite boundary between the new territory III and the old territory I. It began at the west fence of lot 7 (Fig. 6.1). HM-54 and HF-50 moved slowly eastward across our lot almost to the east fence. A few feet to the north, HM-48 traveled parallel to them. There was no singing, and no obvious antagonism in their actions. They refrained from intermingling and they moved quietly. Then the first two flew southward; HM-48 continued foraging eastward into the middle of lot 8.

Eggs were laid in nest 6AJ by HF-49 about 15 March. HF-50, after various difficulties, appropriated her mate's roosting nest, 19C, and laid her first egg on 26 March. In the course of the next 3 weeks we noted no further activities along the boundary. Apparently the chores of nest-building, incubation, and, later, the feeding of nestlings, confined the birds more to the vicinity of their nests. Our attention also was focused more upon the readily visible and accessible breeding nest in our front yard. We may have missed some of the events farther north. In April, we occasionally heard the scratchy call note from that direction. On the evening of 23 April, HM-48 and HF-49 brought their fledglings into lot 7 and put them to bed in nest 22H, the former roosting nest of HM-48. In December, this nest had belonged to HF-49. Nest 22H was only 130 feet (39.6 m) north of nest 19C. Boundary disputes were now noted frequently. In fact, these disputes took place with such regularity that we suspect they were of daily occurrence until 29 February of the following year, when HM-54 disappeared. We observed 13 of these disputes in October, 8 in November (a normally quite month), and 6 in December. We probably missed many more, for our time for observation was limited in the fall.

The invisible boundary line between the two territories was best observed in lot 7, where most of the disputes occurred. It was a bare strip of ground, less than 10 feet (3 m) wide, which extended across the lot between the widely spaced creosote bushes. To the Cactus Wrens it may have been better defined by the conspicuous landmarks along its length, such as cholla cacti. At the west fence the "line" entered a chicken yard and curved abruptly southward between the house and garage in the front of lot 6. In lot 8 it seemed to vanish at the middle of the lot. There were no territorial singing stations close to the boundary line. Such stations were frequently as much as 100 to 300 feet (30 to 90 m) distant. A song at the boundary itself was usually the signal for action.

The north pair, HM-48 and HF-49, had the best view of the boundary. Our house, situated between the breeding nest of HM-54 and HF-50 and the boundary, obstructed the view of the south pair of wrens. Consequently most disputes began when the north pair discovered what appeared to be an impending invasion of its territory. First we heard a song or two from the rear of lot 7. Immediately thereafter came a series of scratchy *scri* notes, as HM-48 and HF-49 flew southward to the "line." As they landed on the ground opposite HM-54 and HF-50, the *scri* calls

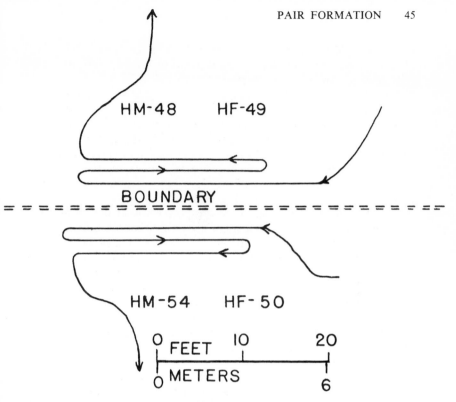

Fig. 5.2   Paths traveled by Cactus Wrens during a boundary dispute on 13 July, 1947, on Kleindale Road, Tucson, Arizona.

increased in number; evidently these calls were uttered by both pairs. Short songs by both of the males and females were interspersed, while the four wrens moved either eastward or westward, in parallel lines, each pair carefully avoiding contact with the other. The group seldom traveled over 15 to 20 feet (4.5 to 6 m) before it reversed its direction and moved back. Occasionally one of the wrens fluffed out its feathers, lowered and spread its tail, and then turned its body in a quarter or half circle. Sometimes it moved sideways, its toes pointing toward the boundary, its head facing in the direction of movement. Frequently the dispute ended after a minute or two of this stereotyped ritual. Then the wrens separated and retreated to the more distant parts of their territories (Fig. 5.2). A few times we noted that several disputes occurred in quick succession, about 50 feet (15 m) apart, as the wrens moved across lot 7 into lot 8.

Now and then the excitement of the ritual increased to the point of breaking out into active chasing and fighting. A chase was always short. If HM-54 chased HM-48 north for 4 or 5 feet (1.2 to 1.5 m), the latter quickly turned about and chased his opponent back. Thus it went, like a miniature seesaw battle, neither side venturing into the other's domain.

A male chased male or female, and was in turn driven back by male or female. Those who gave ground were threatened; those who turned and fought were apparently acknowledged the winners for the moment. Physical combat, when it occurred, was vigorous and vicious. As two wrens came to blows, they rose upward into the air for a foot or more, facing each other, pecking and fluttering until they dropped to the ground again. There they sometimes squirmed in a tangled heap, as they grasped a foot or a wing with their bills and claws. Then they often squealed in pain or fright before they were able to break apart and return to their former positions at the edge of the boundary. In a few moments the battle was over. It always ended in a draw.

Although the fledglings of the first brood could not at first have been aware of the invisible boundary, they nevertheless observed it with remarkable consistency. No doubt their close association with their parents for the purpose of obtaining food kept them for a month or so in their proper territory. Only on a few occasions did these immature wrens stray across the boundary. On 25 May, HM-54 chased a fledgling from territory I northward on our rear lawn. On 16 July, juvenile H-51, of territory III, now entirely independent of its parents, was intercepted on the boundary by HF-49, who climbed on top of the crouching, immature wren and pecked it. The other wrens which had gathered in the course of this dispute remained aloof.

By the middle of July, both families were taking part in these disputes. At first the fledglings stayed in the background, apparently attracted only from curiosity. Later they followed their parents about in the ritual. We feel certain that these young, inexperienced wrens often precipitated the quarrels at the boundary. They revealed their presence there by the noisy, practiced imitation of all the adult sounds. At close quarters they were more aggressive. While the adults were usually content to leave the dispute on a vocal level, the immature wrens often dashed forward to attack. The result was always an increase in excitement. Once when HM-48 was being followed by his fledgling, the male reversed his direction so abruptly that he stepped on the neck of the latter and ran over it. By midsummer most of the immature Cactus Wrens had disappeared. H-52 of territory III remained until December and often participated in the autumn disputes. The lone survivor of territory I, a noband, was also present. Up to November we identified from four to five wrens in every dispute.

Displacement or irrelevant behavior was of frequent occurrence toward the end of each of these boundary disputes. The indeterminate outcome of the conflicts resulted in the wrens running about and picking up nest materials such as chicken feathers and grass stems. Then, as the quarrel ended, the materials were carried to their roosting nests. Rarely, following a dispute, other species of birds were threatened. For example, on 1 July, HM-54 sang from the crossarm of the electric pole on Kleindale Road, just south of lot 7. HM-48 answered from the northeast. Suddenly both wrens began uttering the *scri* note. HM-54 flew north about 200 feet (60 m) to a small pole in lot 8, near the territorial boundary. At once HM-48 arrived from the north and landed on the pole, forcing the other to fly down. Then the wrens faced each other on the ground, 5 feet apart, and moved slowly eastward, while they uttered the *scri* note. HM-48 appeared to be the most nervous of the two. He tried to pull loose some of the dry grasses near him, but failed. HM-54 then abandoned the "line" and returned to his pole on Kleindale Road, where he found that a male Pyrrhuloxia (*Pyrrhuloxia sinuata*) had taken over his singing station. While the Pyrrhuloxia sang, HM-54 moved closer and closer until the former became alarmed and flew; the wren then flew down. Shortly afterward, a Gila Woodpecker lit on the side of the crossarm and uttered its peculiar, sharp, rolling note. HM-54 came flying back at once. He twisted and turned, retreated and advanced, and gradually edged closer. Suddenly he dashed directly toward the woodpecker. The latter backed up to the pole and spread its wings in a defensive gesture and then climbed back upon the crossarm. Again the wren fidgeted closer. This time the woodpecker gave up and flew. Normally a Pyrrhuloxia is ignored and a Gila Woodpecker is deferred to.

Extensive trespassing upon the other's territory was seldom observed. The north pair of wrens occasionally visited the rear lawn with its feeding table and bird bath, but only when the owners of the territory were absent. HM-54 once explored the north end of lots 6 and 7, deep into territory I, and scanned the horizon from a high perch. Since he refrained from singing, he was unnoticed and unchallenged. HF-50, who accompanied him, was more timid and ventured only half as far. That this boundary line was very real was demonstrated on 10 July. At the conclusion of a dispute, HM-48 and HF-49 flew north to the fence of lot 7 and *buzz*ed at a cat which was slowly walking by. HM-54, HF-50, and their offspring H-51, attracted by the danger sounds, flew north into cholla 3, which was about 25 feet (7.5 m) inside of territory I. They perched there quietly, evidently watching and listening, but apparently reluctant to advance and take part in the mobbing of the cat. Soon, however, HF-49 left the cat and returned. She flew directly to cholla 3. The south pair retreated at once to its territory, leaving H-51, who was, perhaps, not so aware of its trespassing. Then HF-49 flew toward the juvenile wren. The latter quickly joined its parents and all was quiet again.

The function of territory. The Cactus Wren's territory belongs in type A of Nice (1941: 458), in which the territory is used for mating and nesting and as a feeding ground for the young. In addition, the territory is retained, with some relaxation of defense, throughout the winter as a feeding and roosting area. We found it to be primarily the property of the male; the female's rights were secondary. Although she assisted in the defense of the territory, she did so, vigorously and successfully, only when her mate was present. Without a partner, she was apparently incapable of defending her territory against usurpation by another pair.

This was very evident in 1945, when the male of territory I was lost. The pair of Cactus Wrens in the adjacent territory III expanded its activities at once into the entire block, with complete disregard for the presence of the widowed female. She offered no resistance, and she later disappeared.

The problem of the critical function and significance of the Cactus Wren's territory remains unsolved. Territorialism is readily demonstrated, and its various elements are easily discerned. We do not know, however, exactly how this form of behavior serves the species. Certain advantages accrue from it, but these advantages are also obtained in some degree by other species through other means. In her survey of territory, Nice (1941: 470) stated that, "The chief function of territory is defense — defense of the individual, the pair, the nest and young. In many cases it also serves to bring the pair together and to strengthen the bond between them."

The maintenance of a territorial boundary by Cactus Wrens did, of course, prevent intrusion and possible interference from other wrens. The question arises, would there have been interference in nest-building, egg-laying, incubation, and feeding of nestlings if no defended boundary existed? In this connection, we must point out that most of the boundary disputes occurred after the nestlings of the first brood were fledged. One must conclude that the Cactus Wrens were so busy raising a family that no time was available for quarrels over a territorial boundary. Somehow, that boundary during the early part of the nesting period was tacitly accepted by both pairs.

There seems little doubt that territorialism assists in pair formation by bringing the sexes together. Yet, all that is required here is that the male make himself conspicuous. This could be accomplished, just as successfully, in a common feeding and nesting area. It appears more probable that territorialism is most useful in maintaining the pair bond, for the restriction of the female to the vicinity of one male must surely be of pairing value.

Territorial behavior in the Kleindale Road area limited the number of pairs occupying the 10-acre (4 ha) tract. Here, too, the problem is left unsolved. Although competition reduced the size of the territories in 1947, nothing is known of what would have happened had there been more Cactus Wrens drawing upon the limited food supply and nesting sites. If Cactus Wrens acquired and defended territories in order to reserve an adequate food supply, then these territories should be larger in the winter months, when insects are less numerous. This appears to be the case; but the wren population per territory also increased, so that nothing was gained until the late winter environmental attrition had removed the surplus offspring.

Perhaps the simplest way of regarding this troublesome problem of the function of territory is to postulate like Wheeler (Nice, 1941: 468) that the basic instinct of self-preservation is equivalent to dominance. It manifests itself not only in the individual's defense of self, but also in an inherent attitude of dominance toward individuals of its own species. This dominance can best be expressed by the male Cactus Wren through maintaining ownership of the small area in which it feeds and sleeps. The female's self-assertion appears to be weaker. Our banding operations were not of sufficient extent to determine if the female held a territory of her own before she found a mate. She probably defends her roosting area against other females. The orderly sequence of the nesting cycle, the perpetuation of the species, and the basic food resources are assured by the male's intolerance of other wrens at this time. The female contributes to a successful cycle by making sure she has no distracting competitors in her habitat.

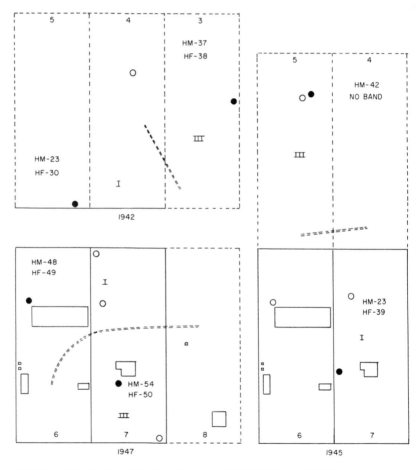

Fig. 6.1. Territories of Cactus Wrens in 1942, 1945, and 1947,
in the Kleindale Road, Tucson, Arizona, area. Open circles indicate
roosting nests; solid circles, breeding nests; and double dashed lines,
assumed territorial boundaries.

6. The Territories From Year to Year

Kleindale Road. The number of breeding territories in the Kleindale Road neighborhood, including those contiguous to the study area (Fig. 2.1), varied from one to five. The local area about our home has been arbitrarily numbered territory I, regardless of the pair of wrens which inhabited it. Beginning in 1939, we noted that the two 1-acre lots east of Edith Street were occupied consistently by a different pair of wrens. These wrens had probably been there, unnoticed, for some years before. The lots were fenced, and the original vegetation was undisturbed. Under such favorable conditions, it is not surprising that Cactus Wrens have occupied this eastern territory (number II) to the present time. It was not always practicable to explore and define the other territories thoroughly. We could determine the location of the common boundaries with fair accuracy, but the extent of these territories could only be estimated. Our early incomplete records show that a pair of wrens nested in lot 6 in 1932, in lot 7 in 1933, in lot 8 in 1934, in lot 6 in 1935, in lot 7 in 1936, and in lot 6 in 1939. We lack nesting data for 1930, 1931, and 1937, but the wrens were present and undoubtedly bred in the area.

In the course of the 8 years, beginning in 1940, the number of territories in the Kleindale Road neighborhood varied as follows: 3 in 1940, 2 in 1941, 5 in 1942, 2 in 1943, 4 in 1944, 3 in 1945, 2 in 1946, and 4 in 1947. In only three of these years, 1942, 1945, and 1947, did territory I suffer encroachment from an adjacent pair of Cactus Wrens. This additional territory we have designated number III for each of these years, although its location changed each year (Fig. 6.1). In the other 5 years territory I was maintained at its usual 10-acre (4 ha) dimensions. The northern peripheral territories were actually marginal in value. They did not contain enough cholla cacti for expansion of nesting facilities. In addition, the chollas were smaller. The hinterland included the dense, brushy creek border, and the sandy, unsuitable bed of the creek. Nevertheless, this area apparently provided room, even if only temporarily, for surplus individuals from other territories.

After 1947, territorial fluctuations smoothed out. The destruction by a sand and gravel company of all the vegetation on the river bank eliminated that habitat entirely. Finally, territory I, our main study area, also suffered severely. Half of its cholla cacti were lost when lots 2, 3, 4, and 5 were scraped clean of all plant life. Only territory II, east of Edith Street, remained intact. The available nesting sites in cholla cacti in territory I were now restricted to lots 6 and 7. From 1948 to 1970, only one pair of Cactus Wrens was able to hold its own in this territory. Surplus wrens from territory II may have assisted in maintaining the back lot population.

Santa Rita Experimental Range. From our experience on Kleindale Road, we knew that adult Cactus Wrens roost alone; unoccupied nests soon become flattened at the entrance. The number of nests in good condition at the beginning of January is a fairly satisfactory indication of the size of the winter population. We assumed that, by the first of April on the Range, the number of remaining nests would house the potential breeding population. By October, the old breeding nests, no longer in use, would deteriorate. New winter roosting nests built by the surviving adults and immature wrens then give a rough approximation of the success of the breeding season.

Time was not available for determining the extent of each territory by watching the individual birds. We have assumed that each pair claimed the land halfway to its neighbors' breeding nests. Cactus Wrens' nests occur in groups, separated by varying distances from other groups. The nucleus of an ideal group is a pair of roosting nests. Several old, weathered remains of abandoned nests can usually be found in the vicinity.

1953. The year began with 41 usable roosting nests, located in 16 well-defined groups. The presence of many old nests suggested that 1952 had been a good year. A few isolated nests could not be assigned to any particular territory. Several nests in territories 7, 9, and 11 (Fig. 6.2), which were at or just outside the boundary lines of the research area, have been included because breeding nests were built later in the same locations. A considerable portion of these territories evidently extended inside the research area. We found several breeding nests under construction on 1 April, but an equal number of roosting nests had been abandoned. The total was still 41 nests. The wrens constructed 16 breeding nests in April (Fig. 6.2). Another nest, found in June, may have been a second brood, or the interdigitation of another territory. The 16 pairs of wrens occupied an area of approximately 76 acres (30.4 ha). We estimate the average size of a territory at 4.75 acres (1.9 ha), the minimum 2.9 acres (1.16 ha), and the maximum 6.9 acres (2.76 ha).

1954. At the end of March, we counted 27 roosting nests. In April and May, the Cactus Wrens built only five breeding nests (territories 1, 9, 12, 15, and 17). A male sang persistently in territory 3, but did not attract a mate. The other territories were quiet; a few nests begun in early

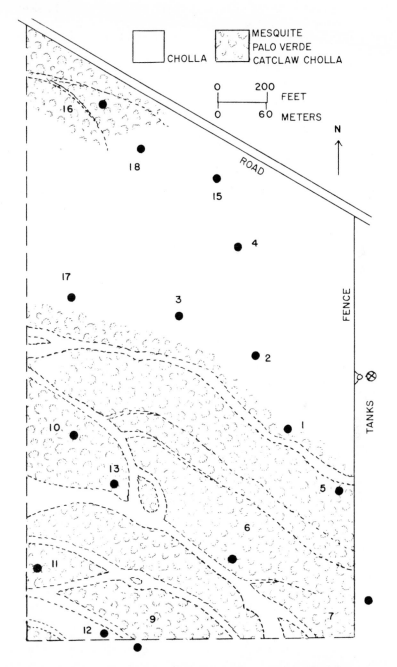

Fig. 6.2. Territories and first brood nests in 1953 in Pasture 5 of the
Santa Rita Experimental Range, 35 miles (56 km) south of Tucson, Arizona.
Solid circles and numerals indicate breeding nests.

spring had been abandoned. Later the five pairs of wrens built nine more nests to house their second or third broods. No estimate of territorial size can be ventured, for each pair probably had more space available than it required.

1955. On 1 April, we counted 21 nests. Again the wrens built five breeding nests (territories 3, 9, 11, 15, and 16). Two of these failed and second attempts were made. Only one pair raised a second brood.

1956. By 1 April, the number of nests had dropped to 18. However, a comeback seemed to be underway, for eight pairs of Cactus Wrens established territories and constructed breeding nests (territories 1, 2, 5, 6, 10, 12, 15, and 17).

Saguaro National Monument. The pattern of nest building by the Cactus Wrens was similar to that observed earlier in suburban Tucson and on the Range. Judging from the number of groups of roosting nests which were located in the late summer of 1962, we assume that only seven pairs of Cactus Wrens occupied territories that year. In the spring of 1963, there were nine first brood attempts; in 1964, there were ten, with another territory active later, possibly a second brood. The total increased spectacularly to 20 in 1965. It then dropped to nine in 1966, but rose in 1967 to 16, and remained at 16 in 1968 (Fig. 6.3). The territories varied from 1 to 4 acres (.4 to 1.6 ha) in extent and were irregular in shape. The average of 13.3 territories per year in the 49 acres (19.6 ha) apportions approximately 3.7 acres (1.5 ha) to a territory. This is the available space. It is a misleading quantity, for at least 4 acres (probably marginal, because of the few chollas in the predominately creosote bush association) along Freeman Road were seldom visited by the Cactus Wrens. Two quadrats in the southeast corner appeared vacant also until 1967 and 1968. Furthermore, in the territories near the borders, some of the birds undoubtedly foraged a bit beyond our arbitrary and invisible boundaries. The 1965 average of about 2.4 acres (.97 ha) per pair is probably a more accurate estimate. This is about half as large as the estimate on the Santa Rita Range. The single pair in the Kleindale Road tract sometimes extended its activities over a 10-acre (4 ha) area, although it did not defend it all.

With such a sedentary species it could be expected that the locations of the nine territories mapped in 1963 would turn out to be remarkably stable in the course of the following years. Excluding the year 1966, when territories II, III, and VII were vacant, each of these nine territories was always occupied by a breeding pair of Cactus Wrens. This preference for certain areas — all characterized by the presence of cholla cacti suitable for nest sites — was clearly illustrated in the population rebound of 1967, when the wrens again occupied four of the new territories which they had established in 1965, but had abandoned in the population decline of 1966. In those years when the increased population pressure squeezed in additional territories, a contraction of the boundaries of some of the original nine could be inferred from the greater frequency of bound-

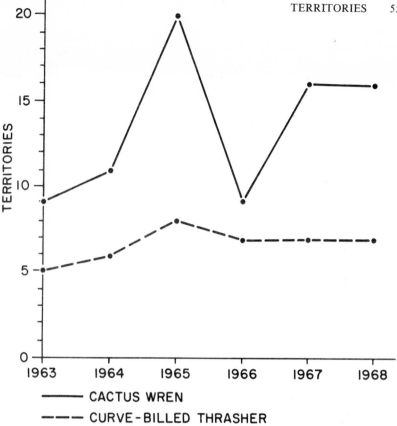

Fig. 6.3.   Saguaro National Monument, Arizona. Number of territories of the Cactus Wren and Curve-billed Thrasher from 1963 to 1968.

ary disputes. However, peripheral habitat in several areas, infrequently utilized at the lower population levels, was readily available without serious conflict. We believe that four more pairs of Cactus Wrens — 24 in all — could have maintained themselves in the 49 acres in 1965 without being forced into unsuitable habitat, but this is mere speculation. Other workers have discovered surprising, inexplicable increases in supposedly stable populations.

Figs. 6.4 to 6.9 show the locations of breeding nests of the Cactus Wrens in the course of the 6-year study, with numerals indicating the consecutive clutches or breeding attempts. Territorial boundaries can be assumed to be roughly half the distance between each pair's breeding nest. The contiguous, extralimital territories, most numerous from 1966 to 1968 along the north and south boundaries, are not shown.

Details of individual changes in breeding areas, and replacements and dispersal in the population, are elaborated in Chapter 17.

Fig. 6.5.   1964 Cactus Wren breeding nests, Saguaro Nationa
Monument. Territory XI did not become active until June
the first nesting attempt may have been extralimita

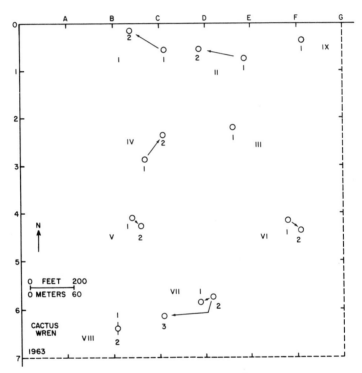

Fig. 6.4.   1963 Cactus Wren breeding nests, Saguaro
National Monument. Circles with Arabic numerals indicate
successive breeding attempts; Roman numerals designate
territories. The wrens in VIII used the same nest twice.
Territories I and VIII were vacant in 1962.

Fig. 6.6.   1965 Cactus Wren breeding nest
Saguaro National Monument. In territory XX the male bu
three nests, but failed to attract a mate. The wrens
territory XV raised their second brood north of the bounda
and returned with their fledgling

[56]

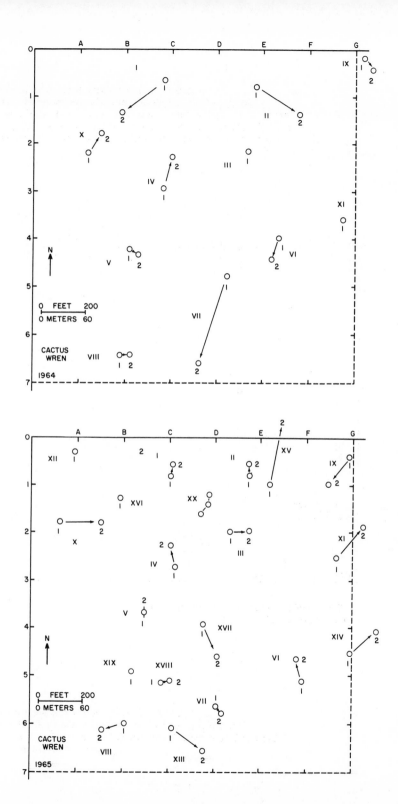

Fig. 6.8.   1967 Cactus Wren breeding nests,
Saguaro National Monument. The two banded wrens in
territory XXI, one a male nestling from XIII in 1966, the female
a 1966 nestling from the extralimital territory south of quadrat
F7, abandoned the territory and possibly nested extralimitally.

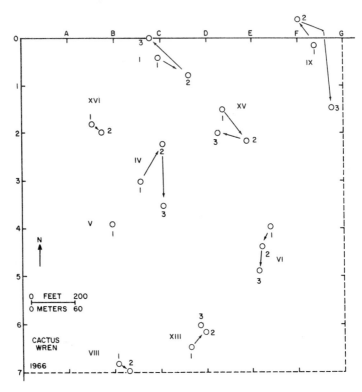

Fig. 6.7.   1966 Cactus Wren breeding nests,
Saguaro National Monument.

Fig. 6.9.   1968 Cactus Wren breeding nests,
Saguaro National Monument. The wrens in XV abandoned
the territory without laying. In XXI the pair which disappeared
in 1967 returned to breed.

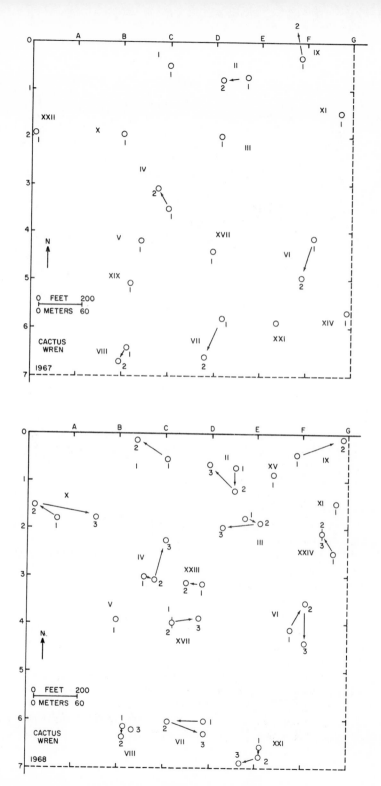

7. The Breeding Nest

Locations

The roosting nests, including those occupied by the adults in November and December, seldom remained intact until the breeding nest was started in the spring. The instability of the Cactus Wrens' nest situation must, at least to the wrens, have been most disturbing. To the observers who tried to follow the course of the winter activities, it was equally disturbing. Nothing seemed to be permanent or settled. Following is a summary of some of the winter-to-spring nesting sequences (for most of these nest locations see Fig. 3.3).

Kleindale Road. 1938-1939. HM-1 roosted in nest 28A, and HF-2 roosted in 27A. The only other roosting nest, 5F in lot 7, was destroyed in February. They built their breeding nest, 28B, in cholla 28, about 2 feet (.6 m) from the male's roosting nest. This was 140 feet (42 m) from the female's winter nest.

1939-1940. Both adults roosted in cholla 6 near Flanwill Street, HM-1 in nest 6J and HF-2 in nest 6I. Just before nest 6J fell to the ground in the latter part of November, HM-1 had rebuilt an old nest 6H in which he continued to roost. It is hard to believe that the wren anticipated the loss of nest 6J; the nest, when it began slipping downward into a vertical position, may have caused difficulty in entry or discomfort in roosting. By the middle of January, the remaining roosting nest, 4C in lot 7, had been damaged. They constructed their breeding nest, 6M, almost midway between the two roosting nests in cholla 6.

1940-1941. The death of HM-1 in 1940 and the disappearance of HF-2 in January of 1941 created considerable uncertainty for us. Most of the rather numerous roosting nests were in lot 6; their tenants were not known. The breeding nest, 35B, of the new pair of Cactus Wrens, HM-23 and HF-22, was located 300 feet (91.4 m) north of cholla 6 in lot 5, close to Flanwill Street. This was the first breeding nest discovered in the northern tier of lots along Greenlee Street since our studies began in 1932. Whether our persistent efforts, sometimes unsuccessful, to capture the wrens in lot 6 for banding purposes annoyed the wrens, or whether the new pair preferred the more open surroundings of the northern portion of the tract, is hard to determine. Later observations lead us to believe that the Cactus Wrens are not easily driven away by ordinary human activities.

1941-1942. Although seven banded wrens were present during the winter, they had only four roosting nests available at the beginning of 1942; all others had been destroyed. When the wrens paired up in the spring and selected their territories, new roosting nests were built near the breeding nest location. HM-37 and HF-38 settled in lots 2 and 3, and built their breeding nest in lot 3; HM-23 and HF-30 chose the south end of lot 5.

1943-1944. At the beginning of 1944, cholla 6 contained four roosting nests; HM-23 and HF-39 had occupied two of them during the preceding months. On 6 January, three of these nests were completely destroyed. HF-39 remained in nest 6W, but this nest, too, was found destroyed on the 15th, and the female began another nest, 6AA, in the same cholla. Eight days later, it also was torn apart. Then the male, who had probably been roosting on the west side of Flanwill Street since the first part of the month, began building nest 6AB; this later became the breeding nest. Another nest, 6Y, was rebuilt, also, we believe, by HF-39. In spite of the destruction of four roosting nests and of the replacement nest 6AA, the pair of Cactus Wrens succeeded in building their breeding nests in the same cholla.

1944-1945. Again cholla 6 provided a roosting nest location for a female, HF-39. Another nest close by, for some reason, was left vacant. The male, HM-23, roosted in lot 7. Suddenly on 6 January, the female began building a new nest, 14C, in lot 7. This nest was abandoned before it was completed. Then on 21 January, she started nest 23F and had it practically finished on the 25th. Now her nest was only 8 feet (2.4 m) from the roosting nest of the male in cholla 22. Two days later, both adults began the task of completing the first nest, 14C. This became the breeding nest. Meanwhile HM-42, who roosted in lot 6 during the latter part of 1944, was forced to find another location when his cholla was cut down on 14 January. (It would doubtless soon have been driven out of territory I, anyway.) He moved north into lot 5, where a breeding nest and a roosting nest were built.

1946-1947. The roosting nests of HM-48 and HF-49 were about 50 feet (15 m) apart at the north end of lot 7. On 27 December, the male's nest was found to be torn apart. Another nest blew down. The female began building nest 19C, south of our house in lot 7, on 21 January. She did not finish it, and evidently continued to roost in her old nest. A month later two noband wrens moved in; one of them took over nest 19C and added more material. HM-48 and HF-49 retreated to the north and west. On 8 March, the female roosted in a new nest in cholla 6, 6AJ. This nest became the breeding nest. We discovered the male roosting in the female's old nest in lot 7.

Both of the newcomers worked on nest 19C in February, but the nest was damaged by a Curve-billed Thrasher. Nest 8P in lot 8, about 100 feet (30 m) east of 19C, was started. Work continued on nest 19C, and it was occupied at night by the male. The female HF-50 (now banded) worked on nest 8P and roosted in it until 18 March, when the nest was destroyed. The following day, the male began work on nest 15C in the southeast corner of lot 7. Then both adults joined in finishing nest 19C, which had now become the female's roosting nest. The male spent the nights at least as far east as Edith Street. On 22 March, the pair shifted their labors to nest 15C; then they went to work again on nest 19C. This finally became their breeding nest.

1947-1948. The division of the tract into two territories was maintained throughout the winter. For the third time in our studies, a male and a female Cactus Wren of territory I roosted in the ever-popular cholla 6. HM-48 roosted first in nest 6AN; when it fell to the ground, he built nest 67B at the west edge of lot 6, 45 feet (13.5 m) southwest of cholla 6. Then, at the end of December, he began work on nest 6AO. The roosting nests of HM-54 and HF-50 in territory III remained in good condition up to the time the male disappeared about 28 February. The north pair of Cactus Wrens then took over the entire area. HF-49 began work on nest 23J in the north third of lot 7 on 1 March. Two weeks later it was torn apart; only one good nest remained in cholla 6 — 6AK, the female's roosting nest. HM-48 built himself another roosting nest in cholla 67. Finally, unexpectedly, HF-49 laid her eggs in nest 6AK.

1956-1957. Six Cactus Wrens, four of which were banded, were present in November and December of 1956. When the situation "stabilized" in January, after six nests were damaged or destroyed, only two wrens, HM-70 and HF-71, were left in possession of the territory. On 21 January, HF-71 began construction of nest 21C, about 40 feet (12 m) north of our house, and roosted in it. The male roosted in nest 17G in the southwest part of lot 7. This nest was soon destroyed; another was begun; and it, too, was torn apart. Some repairs were attempted, but again the thrashers damaged it. Then the male, apparently without opposition, began to carry nest material to the female's nest 21C. The female started another nest, 21D, a few feet away from her former nest. The Curve-billed Thrashers now turned their destructive activities upon the new location. The female wren held on, roosting regularly in nest 21D, although it had been torn up at least three times. At last, on 26 January, she gave up and began a new nest, 94B, in the northeast corner of lot 7. Then, on 8 February, we discovered that the breeding nest, 93B, was being constructed in a cholla near the west fence of lot 6, some 320 feet

(97.5 m) southwest of her last roosting nest, and 180 feet (54.9 m) from the male's roosting nest.

From the foregoing accounts it might be conjectured that the selection of a nest site was not always a matter of simple choice of a favorable location. Could the competitive thrasher be a determining factor? Such a supposition, we believe, is wrong, for the Curve-billed Thrashers did not destroy the breeding nests of the Cactus Wrens, so far as we could observe. It was the unattended, unprotected roosting nests which suffered. Once a breeding nest was begun, the presence of the Cactus Wrens at the nest seemed to be sufficient to deter the thrashers from their depredations at that particular point. There are some exceptional events, very difficult to explain adequately, in this connection, which will be related later. In a few cases, too, it was not always possible to determine, because of the lateness of the season, if a newly begun nest, which was destroyed, had been intended to serve as a breeding nest or as a roosting nest.

Santa Rita Experimental Range. 1953-1956. The Cactus Wrens chose the periphery of the crown of the cholla cacti for their nest locations. The nests were conspicuous; to us, at least, there appeared to be no attempt at concealment from any enemy. Nests were not always securely anchored among the spiny cholla joints; strong winds evidently blew some of them away. Others, we feel sure, were torn loose by cattle which found the bundle of dry grass attractive. We occasionally saw cattle with cholla joints impaled on their jaws. By far the greater part of the nest destruction was so similar to that which the Curve-billed Thrashers perpetrated at Tucson that we must attribute it to them.

Saguaro National Monument. 1963-1968. The Cactus Wrens constructed about 60 percent of their 6-year total of 154 breeding nests in the 49 acres (19.8 ha) in cholla cacti (Fig. 7.1), preferring the dense, crowned, jumping cholla four to one over the staghorn cholla (Table 7.1). They placed them at heights ranging from 3 to 7.5 feet (0.9 to 2.3 m).

Table 7.1. Cactus Wren breeding nest locations, Saguaro National Monument, 1963 to 1968.

Nests placed in	Number	Percent
Opuntia fulgida	4	} 48.1
Opuntia fulgida var. mammillata	70	
Opuntia versicolor	17	11.0
Opuntia spinosior	1	.6
Cercidium microphyllum	5	3.2
Phoradendron californicum	2	1.3
Carnegiea gigantea — live	48	31.2
Carnegiea gigantea — dead stump	5	3.3
Carnegiea gigantea — in hole	2	1.3
Total	154	100.0

Fig. 7.1.   Cactus Wren nest in jumping cholla,
Saguaro National Monument, Arizona.

Fig. 7.2.   Cactus Wren nest in palo verde, Saguaro National Monument, Arizona,
not typical; most nests in this tree are placed in dense clumps of mistletoe.

Nests in palo verde trees were from 6.5 to 12 feet (2 to 3.7 m) up (Fig. 7.2). Leafless mistletoe, a common parasite of the palo verde, furnished secure locations for two of the nests. Mesquite trees apparently had too open a structure to support nests.

The remaining 35 percent of the nests were in giant saguaros (Fig. 7.3). All but seven of these nests lay in crotches between the curved branches, at the point where they spread outward and upward from the trunk, usually 10 to 14 feet (3 to 4.3 m) above the ground. The range was 8 to 20 feet (2.4 to 6.1 m). A large woodpecker hole, 18 feet (5.5 m) up, and an irregular, callused hole, 7.5 feet (2.3 m) up, served for two other nests. Five nests were placed 8 to 15 feet (2.4 to 4.6 m) up in saguaro stumps. These dry stumps, averaging 2.5 per quadrat (range none to 10) were the remains of saguaro victims of bacterial necrosis, a disease caused by *Erwinia carnegieana,* prevalent in section 17 of Saguaro National Monument (Alcorn and May, 1962: 156-158). At first most of the heavy, curved branches of the diseased saguaros, and sometimes the top of the trunk, drop to the ground, spreading out like the spokes of a wheel. Finally the rotting, disintegrating skin of the trunk breaks loose and slides to the ground, exposing a cylinder of bare ribs, 6 to 10 inches (15 to 25 cm) in diameter. Inside the cage of ribs, the pulp decays, dries, and gradually escapes through the cracks between the ribs. At the top of the

Fig. 7.3.   Cactus Wren nest in crotch of saguaro,
Saguaro National Monument, Arizona.

Fig. 7.4. Saguaro stump, Saguaro National Monument, Arizona, used for Cactus Wren nest location.

stump the drying ribs fan out. The Cactus Wrens entered the stump, either between the ribs or through the open top, to construct their nests upon the pulpy interior (Fig. 7.4).

Nests seldom contained many feathers, in contrast to those in the earlier Kleindale Road study, where chicken feathers were always available. Occasionally the wrens gathered the felt-like material found among the spines at the ends of saguaro branches.

Entrance Orientation

Since the Cactus Wren occurs in many desert situations from southern California east to Texas and southward into Mexico, its nest locations naturally vary with the type of thorny vegetation present. Woods (in Bent, 1948: 219-223) lists a large variety of desert plants which the wrens have found suitable for nest sites. In addition, in the outskirts of towns, introduced plants such as palm and olive trees are sometimes chosen.

Differences in the configuration of nesting sites, their accessibility, and the practice of facing the entrance of the nest outward, are evidently factors which influence the orientation of the entrance. Table 7.2 lists the orientation of the nest entrance of nine first broods and 33 later broods in the Kleindale Road area. Table 7.3 of the Santa Rita Experimental Range lists all broods together; we were not always sure of their sequence.

In the Saguaro National Monument tabulation, Table 7.4, the first and later broods are separated. This covers the entire 6-year period, and includes nests in adjacent extralimital territories.

Since winds here show little variation in annual pattern, no year-to-year variation in nest orientation should be expected. We find none in our samples. We feel that as the number of samples of nest orientation increase the curve will smooth out and flatten. Ricklefs and Hainsworth (1969: 35), apparently using a sample of only 63 breeding nests, 27 of them "early" and 36 of them "late," endeavor to show in their histograms that, in nests in March and April, directions between west-southwest and north were "significantly" avoided and that in "later" nests, May and June, southwesterly directions were "significantly" favored. Even though such a statistical analysis may reveal that the observed orientations fit a normal distribution curve, it does not explain the large number of variations in orientation at each side of the peak. Neither does it prove that wind direction and velocity were of prime importance, or of any importance, in the determination of the entrance orientation. Other factors could be involved.

Table 7.2. Orientation of Cactus Wren breeding nest entrance on Kleindale Road, Tucson.

N	NE	E	SE	S	SW	W	NW	Total
				First brood				
1				3	3		2	9
				Later broods				
3	4	5	2	9	1	3	6	33
				Total				
4	4	5	2	12	4	3	8	42

Table 7.3. Orientation of Cactus Wren breeding nest entrance on Santa Rita Experimental Range, Arizona.

N	NE	E	SE	S	SW	W	NW	Total
6	3	4	4	8	5	7	11	48

Table 7.4. Orientation of Cactus Wren breeding nest entrance in Saguaro National Monument, Arizona.

N	NE	E	SE	S	SW	W	NW	Total
				First brood				
18	9	14	6	19	14	12	8	100
				Later broods				
12	4	13	13	10	8	17	13	90
				Total				
30	13	27	19	29	22	29	21	190

We are not aware of evidence indicating that March-April prevailing southwest winds at afternoon temperatures of 70 to 80F (21.1 to 26.6C) cause such great discomfort to an incubating Cactus Wren or its nestlings that it induces it to avoid a southwest entrance to its nest. The greatest discomfort should occur at the minimum temperature just before dawn when wind is usually absent. In regard to the "late" nests, presumably orientated to take advantage of the prevailing southwest winds in hot afternoons, moving air at 90 to 105F (32.2 to 40.6C) or higher can have little cooling effect. Furthermore, the quantity of air moving through a nest depends not upon the direction of the wind, but upon its force and the frictional resistance it encounters. The Cactus Wren's nest is not a one-way valve. The lattice of grasses, thin at the entrance, gradually thickens along the vestibule, and becomes a densely packed wall of additional materials surrounding the bowl or cavity of the nest. A column of air in the vestibule, forced inward by outside pressure, finally escapes through the permeable upper parts of the nest. Wind from the opposite direction encounters the same resistance as it strikes the rear of the nest. It diffuses into the inner end of the vestibule and leaves through the entrance. Both air currents are modified slightly by swirling effects around the nest. The air entering from the rear can only be kept out of the vestibule by a greater internal pressure — an impossible condition. The most obvious disadvantage of an entrance facing south, southwest, or west — that is, into the late afternoon sun — would be the heat. Nestlings, which in the final week of their nest life, often crawl into the open entrance to await food, would be subjected to the direct, intense heat of the sun, and forced to retreat to huddle tightly in the poorly ventilated bowl of the nest. Evidently this is not a serious problem, for over one-third of the nests in both categories faced southerly directions.

Further evidence of the unpredictability or the random orientation of breeding nests comes from the choices made for second and third clutches. Eight pairs chose to orient their second nests similar to their first. Some used the same nest for the second brood. Most of those which laid triple clutches varied the construction with three different directions; and the lone four-clutch pair laid the first clutch in a cholla, entrance east, the second in a saguaro stump, entrance southwest, the third in another saguaro stump, entrance northeast, and for the fourth it chose the first stump again with its southwest entrance. The orientation of the breeding nests of year-old wrens was completely random. Often it was directly opposite to that of the nest from which they had been fledged the year before. Nest locations, too, were equally random; only the form of the nest appeared to be inherited.

The importance of the saguaro is demonstrated in Table 7.5, which reveals an unexpected and inexplicable preference for saguaros for breeding nest locations from the first to the second nest in double nestings, and from the first to the second and third nests in triple nestings. Palo Verde trees of considerable height were always available, had the Cactus Wrens

Table 7.5. Change in Cactus Wren nest locations in Saguaro National Monument, Arizona.

Type	Clutch	Cholla (Opuntia)	Saguaro (Carnegiea)	Palo Verde (Cercidium)	Total	Percent in Saguaros
		Nest placed in				
Single		13	6		19	31.6
Double	1	38	4		42	9.5
	2	12	25	5	42	59.5
Triple	1	17			17	0.0
	2	9	7	1	17	41.2
	3	3	13	1	17	76.5
Total		92	55	7	154	35.7

desired nest locations higher than those in cholla cacti, but only seven such trees were used. Frequent inspections of nests for eggs, and removal of nestlings for banding, possibly contributed to the selection of apparently safer locations. Some wrens, however, laid two consecutive clutches in nests in saguaros. In a study area east of Tucson, Ricklefs and Hainsworth (1969: 35) reported a very slight upward shift in nest height in cholla cacti. "Late breeding nests" averaged .4 foot (.12 m) higher than "early breeding nests." They mentioned no nests in saguaros.

We visited the Santa Rita Range usually in the mornings; winds at that time in the spring were light. On some afternoons the winds were southerly. Although a wren would doubtless be more comfortable facing the wind, so far we see no evidence that the wind direction is of any importance in the position of the nest entrance. Nest construction usually begins in the early morning when wind is absent or very light. The location of the entrance is decided upon at the very start, for it faces the direction from which the wrens bring the material into the nest. In the Rillito Creek area and also in the Saguaro National Monument, the wind is usually from the east in the morning, and gentle. It gradually swings south and then west, and by mid-afternoon it has reached northwest, where it remains, somewhat stronger, until dusk. Heavy, dusty winds in the spring may be either from the east or west, and they often last 2 or 3 days and nights. Nests are not begun at this time, for the material blows away.

Site Selection

We had no better success in observing the beginning of the first breeding nest of the season than we had in seeing the beginning of the roosting nests. The choice of a particular site, however, was probably made by the female. Sometimes, early in the year, we found a pair of Cactus Wrens prowling through the branches of a cholla cactus, apparently examining it for a suitable nest location, instead of searching the spiny joints for insect food. Occasionally the male would be carrying a grass stem in his bill as he followed his mate or moved from one side of her to the other. Usually his movements were far from deliberate; he appeared excited and eager and jumped from joint to joint with quick, abrupt twists of the body. Then with outstretched neck he peered into the tangle of branches while he held the nest material. Some of the rapid movements reminded one of the courting behavior of the ever-present male House Sparrows, but the elaborate posturing and noisy sounds were absent. The inspection of the cholla always ended, while we were present, in the departure of the wrens to another portion of the tract.

The events of the spring of 1947, previously mentioned, suggest that the female makes the final choice of a nest site, and it is sometimes at the expense of the male. The destruction of HF-50's roosting nest, 8P (possibly also her intended breeding nest), on 18 March, was followed quickly by her appropriation of the male's partly finished nest, 19C. Then, when HM-54 built another roosting nest, 15C, for himself, he also assisted his mate in the completion of her nest. Suddenly, for no apparent reason, at least to us, the female began carrying material to the male's new roosting nest. Finally both wrens went back to work on nest 19C. It is possible that at first HF-50 felt that she was being disturbed too often, since her nest was so close to our house. The male's nest in the southeast corner of lot 7 was farther away, but it had disadvantages, too. The Kleindale Road traffic was only a few feet away.

It would be easy, but certainly misleadingly anthropomorphic, to say that this was a perfectly matched pair of Cactus Wrens that worked together for the common good, at least after the female took possession of the male's roosting nest. The behavior of the male suggests a different explanation. He did not appear eager to be displaced from his roosting nest; in fact, one received an impression of reluctance. On 19 March, the day after the loss of the female's nest, HM-54 uttered a peculiar whining sound at the entrance of nest 19C as though he were disturbed; then he sang. Soon afterward both wrens were inside the nest. The same painful, whining sound or *squeal* was heard again. HM-54 was first to come out of the nest. By midmorning, the male was busily constructing his new roosting nest, 15C. Possibly the presence of the female in his nest was an indication that he must give it up.

The next day, although both adults worked energetically on nest 19C, apparently without friction, the male continued to utter this *squeal*. Several times we observed that he entered the nest without material and *squealed* inside. When he came out, he usually sang once on the doorstep. At 1820, when HF-50 arrived in the cholla, HM-54 slipped quickly into the nest. The female looked in but did not attempt to enter. The male soon came out, for it was still too early to retire for the night. Twenty minutes later, when both wrens flew to the nest, HF-50 was the first to enter. The male peered into the nest for several moments, then left. On 21 March, we noted the same behavior, except that the male followed his mate inside. He remained there only a short while, before flying east to roost. Nest 15C was not occupied as a roost until the following night. We had never heard this *squeal* before, and have very few records of its occurrence in later years. It would appear to be an utterance of protest from the male, mild to be sure, for no quarrel ensued as the female dispossessed the male of his roosting nest. We heard this protest only when the female was near or in the male's nest at the beginning of the breeding season. In some way, the male may have sensed that the nest was to become a breeding nest. Here we tread upon dangerous ground. We do not know how the male is led to help in the construction of a breeding nest, nor do we know how he distinguishes this nest from the recently-built roosting nests. We have observed that the usual protest upon eviction from a roosting nest is an angry sputter of *scri* calls, sometimes followed by a fight. The loser *buzz*es as it retreats to find another roosting place. We did not note the *squeal* at nest 15C, which the male had just begun. Perhaps the attachment to a newly begun nest is not as firm as that to a long-occupied one.

We believe that when the breeding season begins, the period of complete male dominance is over. The female, evidently by means of increased aggressiveness, is able to determine to some extent the course of her own affairs. She can and does at times select her own breeding nest site for the first brood.

Breeding Nest Construction

Both sexes constructed the first breeding nest of the year. Under normal, unhurried, and undisturbed conditions, the division of labor was probably nearly equal. It was not unusual to observe the male and female arriving at the nest at the same time, with bills full of nest material. Sometimes, before the roof was in place, they would enter side by side and deposit the grasses upon the floor, or poke them into the thin, rising walls of the nest cavity. The first to finish left through the open top. Later, after the roof and vestibule were outlined, only one could enter at a time. Occasionally both birds would be working inside, but usually at this stage of construction the visits appeared to be so timed that no waiting was necessary at the entrance.

At first we supposed that two Cactus Wrens working together could build a nest twice as rapidly as one could working alone. This, however, does not appear to be the case, for the inattentive periods were often long and irregular. The rate of construction was probably set by the female. When she left her nest building, the male followed her; and the urgency with which she worked was, no doubt, dictated by the nearness of the egg-laying stage.

There are so many variables that more data would be desirable. If we average the number of visits to roosting nest 23A (Chapter 1) for the three active working periods, we get one visit for each 0.80, 0.88, and 1.1 minutes respectively, or roughly one visit every 0.9 minute. Any faster rate of delivery of materials would leave very little time for their placement in the nest. At breeding nest 28B, in 1939, two wrens working together made ten visits in 28.5 minutes; then, beginning 15 minutes later, they made 11 visits in 20 minutes, the average being one visit in 2.8 and 1.8 minutes. The breeding nest 19C of 1947, observed for 40 minutes, revealed 26 visits with nest material, an average of 1.5 minutes between visits. These averages of elapsed time between visits for two wrens working together are approximately twice as great as for one wren working alone on its roosting nest. We watched the work at roosting nest 23A early in the forenoon in August; we watched breeding nest 28B at noon and 19C in midmorning, both in March. Although nest construction proceeded more rapidly early in the morning, the breeding nest rate did not equal or exceed the roosting nest rate.

No satisfactory method could be found for determining when the first breeding nest of the year was completed, because installation of nest material usually continued up to the time the first egg was laid. In some instances this period of time was not what one could call a normal one. Loss of mates, nest destruction, unfavorable weather conditions possibly, and many human disturbances in the Kleindale Road area, complicated the picture until we have such extremes as 5 to 41 days from the time of beginning of nest construction to the laying of the first egg. The average time in the Kleindale Road area for the 13 years in which we have sufficient data is 16.1 days. If we ignore the abnormal high of 41 days, our average is 14 days. Since visits to the Santa Rita Experimental Range, south of Tucson, could usually be made only at weekly intervals, exact data on nest building was not obtained. At least seven breeding nests in that area, however, required more than 15 days in construction.

The nest of the Song Sparrow is built entirely by the female (Nice, 1937: 94); the stimulus for the initiation of nest building probably arises from her physiological condition. Although the situation in regard to the Cactus Wren is complicated with more variations in behavior, it seems likely that the female normally installs the first nest material when she is ready to begin her first brood of the year. The male then assists in the construction as long as the female is present. A few exceptions to this sequence have been noted in the summaries of winter nests. The female does occasionally take over a nest built by the male for a roosting place. This can occur because of nest destruction, or perhaps from a sudden development of favorable environmental stimuli that makes egg-laying urgent. The variation most difficult to account for is the one in which the female laid her eggs in her own roosting nest. This did not occur often, and in one case the roosting nest was new. It may actually have been intended for a breeding nest. Some nests in neighbors' lots were difficult to watch; we may have missed the work of the male. For her later broods a female did not need to build a nest. The nest was already waiting for her; it had been constructed by the male while she incubated.

8. The Start of Laying

We had hoped in the course of this study that a sufficiently large sample of data on the time of laying of the first egg could be accumulated to enable us to place the results on a statistical basis. This hope proved futile, for seldom was more than one territory available and accessible in the Kleindale Road vicinity during any one year. In addition, the variations in the time of laying from year to year proved to be very great. Had we been content to work on the life history of the Cactus Wren for only a year or two, these variations would have remained undetected, with most of the difficult problems happily unnoticed.

Copulation has been observed as early as 18 days before the first egg was laid. It continues up to at least the laying of the eggs. We have a record of a 40-day interval, but in this instance the wrens experienced difficulties in their nest building, and they were doubtless delayed. This is also our earliest record; it occurred on 12 January 1957. In the extremely early nesting of 2 January 1958, copulation must have taken place in December, even before the female's selection of a breeding nest site.

In our observations, the female indicated her readiness, from an elevated perch, by crouching low and singing a rapid series of *rar rar rar rar rar* notes several times. The tone appeared to be at a somewhat higher pitch than that of the normal song. Her bill pointed slightly upward, and she fluttered or quivered her wings. The male responded by flying to her quickly. Sometimes he spread his tail upon arriving, as he might do in the customary recognition display. Then he too fluttered his wings, just before he mounted the female to make momentary contact. After completing the act, they usually perched quietly a few moments before flying to the ground, but in one observation the male pecked the female on the head. She did not move. Then he pecked her twice more before she left him. We have only one record of the female flying to the male.

This occurred when the male was singing, and he stopped at once when she crouched and fluttered her wings.

Only one good observation on the length of time it takes for the growth of the ovum in the Cactus Wren was made. On 29 May 1958, we discovered that the three 3-day-old nestlings in nest 25H were gone. Two days later, the adults were at work on another nest, and the first egg was laid either on 4 or 5 June. Nice (1943: 210-211) states that the "start of nest building in the Song Sparrow roughly corresponds with the rapid growth of the ovum . . . which normally begins 5 days before the egg is laid." The basis for the latter part of this statement is her observation that after the loss of a set of eggs, the next egg is laid 5 days later. The interval of 5 days in the Song Sparrow and 6 to 7 days in the Cactus Wren occurred not at the beginning of the breeding season, but later, when the female's reproductive condition must have been in an advanced stage of readiness. Our Cactus Wrens, on the average, began their first nests at least a week before the beginning of the 6- to 7-day period of rapid egg development.

In Table 8.1 we give the estimated date of the first egg of the year in the Kleindale Road area from 1934 to 1958. There are 3 years missing, 1937, 1949, and 1951, in which no data were obtained. Although one sample per year is obviously inadequate for any generalization, there are two interesting and suggestive years in the table. In 1942 there were three territories (Fig. 6.1, the Edith Street territory not shown). The date of the first egg was the same in two of them, and only 3 days later in the third. There were two accessible territories in 1947, with the egg laying dates 11 days apart. Had it not been for the destruction of HF-50's first breeding nest in this year, the dates would probably have been closer together. We think the possibility is present that many of the dates in the table may be the normal ones, and should be considered as the time at which a larger population would have begun to lay. If this be true, then some explanation will have to be devised to account for the yearly variations.

We gradually came to believe that for a given species in a uniform habitat, the approximate date of physiological readiness for breeding is probably in part determined genetically, in that it is the result of centuries of selection for response to regular, periodic, seasonal changes. In a desert environment the supply of insects and spiders must be rather low by the end of the colder months. Theoretically the egg laying should be so timed that sufficient food is available for the young when the eggs are hatched. If they are hatched too early, they may suffer for want of the proper food; if they are hatched much later, the time for additional broods is shortened. The dependence of this food supply upon a proper interrelationship of temperature and precipitation must be apparent. Without winter rains, the plants and their accompanying insects will not develop, even though temperatures are favorable. Rainfall without a sustained increase in temperature is likewise inadequate for growth.

Table 8.1. Kleindale Road. Date of first egg, with summary of temperature and rainfall.

Year	Date first egg laid	Average of mean temperatures 7 days before laying[a]	Average of mean temperatures 14 days before laying[a]	Rainfall from 1 October to date of first egg[a]
1934	21 March	65.5 (18.6)	64.6 (18.1)	3.65 (92.7)
1935	23 March	56.1 (13.4)	53.8 (12.1)	7.68 (195.1)
1936	1 March	56.0 (13.3)	55.1 (12.8)	5.01 (127.3)
1938	14 March	60.7 (15.9)	58.4 (14.7)	2.83 (71.9)
1939	16 March	59.3 (15.2)	55.4 (13.0)	3.01 (76.5)
1940	20 Feb.	49.0 (9.4)	48.5 (9.2)	2.08 (52.8)
1941	24 Feb.	59.0 (14.9)	58.6 (14.8)	8.14 (206.8)
1942	7, 7, 10 March	53.7 (12.0)	53.5 (11.9)	5.39 (136.9)
1943	20 March	58.0 (14.4)	57.1 (13.9)	2.44 (62.0)
1944	20 Feb.	47.7 (8.7)	48.6 (9.2)	1.81 (46.0)
1945	8 Feb.	53.0 (11.7)	50.5 (10.3)	5.21 (132.3)
1946	13 March	58.7 (14.8)	57.2 (14.0)	4.35 (110.5)
1947	15, 26 March	53.1 (11.7)	53.3 (11.8)	2.37 (60.2)
1948	13 March	53.0 (11.7)	50.4 (10.2)	3.81 (96.8)
1950	1 March	58.7 (14.8)	60.0 (15.5)	3.21 (81.5)
1952	3 March	55.1 (12.8)	52.9 (11.6)	5.37 (136.4)
1953	23 March	61.7 (16.5)	62.0 (16.7)	4.84 (122.9)
1954	25 Feb.	57.0 (13.9)	57.1 (13.9)	1.21 (30.7)
1955	12 March	64.4 (18.0)	59.5 (15.3)	2.18 (55.4)
1956	7 March	58.8 (14.9)	56.9 (13.8)	2.60 (66.0)
1957	21 Feb.	62.5 (16.9)	63.7 (17.6)	3.05 (77.5)
1958	2 Jan.	54.4 (12.4)	54.4 (12.4)	3.89 (98.8)

[a]Temperature in parentheses is in degrees Centigrade; rainfall in parentheses is in millimeters.

In Table 8.1 we include the average of the mean daily temperatures for the 7 and 14 days preceding the laying of the first egg of the year. These intervals correspond to the presumed time for the rapid development of the ovum and the time required to build the breeding nest, respectively. The temperatures are taken from the University of Arizona weather station situated about 3 miles (4.8 km) southwest of our Kleindale Road study area. The range of variation is slight; it is 17.8F (9.9C) for the 7 days, with extremes of 47.7F (8.7C) and 65.5F (18.6C); it is 16.1F (8.9C) for the 14 days, with extremes of 48.5F (9.2C) and 64.6F (18.1C). There are 15 and 17 of these 22 layings in the 50F (10C) to 60F (15.6C) interval, respectively. Early layings were not necessarily at the lowest mean temperatures. Fifteen of our dates are in March.

Early nestings, we believe, can usually be explained by favorable environmental conditions. Rather difficult to account for is the fact that three females laid earlier in the second year that we had them under observation than they did in the first; possibly they were birds only 1 year of age when first observed. HF-2 laid her first egg on 16 March 1939; the next year the date was 20 February. HF-49 laid her first egg on 15 March 1947; in 1948 she laid an egg on 13 March. There are no data for 1949, but in 1950 she laid a first egg on 1 March. HF-71 laid her first egg of 1957 on 21 February; the following year she laid the first egg on 2 January.

In 13 of the 22 years, laying began after a steady rise in mean temperatures of several days' duration. In most of the other years, a period of above normal temperatures preceded the laying. These temperatures were, of course, preceded by a rise, which may have provided the stimulus toward nest building. The early laying of 21 February 1957 occurred near the bottom of a temperature decline of 6 days, but the previous peak, after a 9-day climb, had reached the unusual extreme of 72F (22.2C). The still earlier record of 2 January 1958 occurred after a 3-week period of above normal temperatures, some of which were as much as 10F (5.6C) above the average for Tucson at this time of year. In 1940 and 1944, the start of laying was at temperatures below normal, but in both these instances an above-normal peak occurred about 2 to 3 weeks previously.

The effect of rainfall in the Kleindale Road area is far more difficult to interpret. If we tabulate the precipitation from 1 October to the date of the first egg (Table 8.1), we obtain 3.65 inches (92.7 mm) for 1934 and 7.68 inches (195.1 mm) for 1935, 2.08 inches (52.8 mm) for 1940 and 8.14 inches (206.8 mm) for 1941, with the corresponding egg dates on 21 and 23 March, and 20 and 24 February. There is no evidence here that laying is postponed in periods of low rainfall until mean temperatures are higher, or that total quantity of rainfall, by itself, influences the time of laying. Normal rainfall in October, November, and December, accompanied by mild temperatures, initiated and accelerated the growth of winter annuals to a very striking degree. For instance, filaree and bladder-pod, which usually begin flowering in the first part of February, were in bloom in the last week in December 1957, and their first shoots were observed as early as the second half of October. Another mustard (*Sisymbrium irio*) and a grass (*Schismus barbatus*) appeared in November. The earliest Cactus Wren egg was laid on 2 January 1958. Whether this was the result of mild temperatures, or the combination of mild temperatures, rainfall, and new fresh vegetation, we cannot determine. Mesquites, creosote bushes, and chollas varied considerably in their time of flowering, but their growth could hardly have had any influence on early egg laying, for nesting was always well under way by the time these larger plants were flowering.

Earlier we have tried to show that our population of Cactus Wrens was composed of immigrants from the neighboring population north of Rillito Creek. The two populations should normally begin laying, if our supposition of genetic similarity is correct, at about the same time. In fact, the entire Tucson valley, although exhibiting considerable variations in vegetational aspect, is rather uniform in winter temperature and rainfall. On Kleindale Road, close to the Rillito Creek trough, the winter night-time temperatures were lower than those on higher ground. We should expect little difference in time of egg laying anywhere in the area. Here again we are plagued with insufficient data for a proper comparison. Working with estimated dates, we have in 1941 the first egg on 2 March, at a point 0.25 mile (.4 km) northwest of our home, still on the south side of the Rillito. The first egg in the home area was on 24 February. Seven miles (11.2 km) to the southeast, the first egg was laid on 6 March. In 1955, 3 miles (4.8 km) to the east, along the edge of the foothills of the Santa Catalina Mountains, bordering the Rillito, eggs were laid on 30 March and 2 April; in our area it was on 12 March. A few more data are at hand from the Saguaro National Monument, 11 miles (17.6 km) to the southeast of our home. In 1954, the first egg was laid on 17 March, estimated from information furnished by Hal Harrison; on Kleindale Road we have 25 February. In 1956 we estimated the first egg was laid on 23 March, as compared with 7 March in our area; in 1958, the first egg was laid on 16 February in one nest and 26 February in another, compared with 2 January in our lot. In the same year we estimated that an immature Cactus Wren collected by Allan R. Phillips 17.5 miles (28 km) southeast of Tucson came from an egg laid about 20 February. Evidently the southeastward population was also influenced by the mild winter. We have but few egg dates from residential parts of Tucson: 8 March and possibly 22 February 1956, estimated from data given to us by John Chalk, and 12 March 1954, by Allan R. Phillips. While we realize that our data are meager and that some of the estimates could be wrong by as much as plus or minus 5 days, we are faced with the peculiar fact that egg laying took place later in all the adjacent areas than it did in our small 10-acre (4 ha) study tract. The available information suggests that egg laying begins earlier in and close to cities. No doubt various new insects multiplied far more abundantly and probably earlier in the well-watered ornamental shrub areas than in the dry sandy expanses among the creosote bushes. Perhaps it is these insects that are the key to the early nestings.

Herbert Brown (1888: 116) reported that "nesting was well under way" in the Tucson region on 13 March 1885; Scott (1888: 162) said that eggs are laid as early as 20 March in the Santa Catalina region. Brandt (1951: 679) stated that fresh eggs could be found in the desert about Tucson as early as 15 March, "but the peak of season is not until a month later; while over in the high country at 5000 feet, near where the San Pedro River crosses the Mexican line, I found a nest with 4 fresh eggs on June 21, 1944." We have purposely omitted the listing of nesting records from other states, because they are too scattered and they lack important data. The practice followed for many years in Bent's "Life Histories of North American Birds" of recording the length of the egg-laying season, and then attempting to establish the most frequent period is misleading. For instance, in the case of the Cactus Wren (Bent, 1948: 231) we have: "Egg dates.-Arizona: 82 records, March 10 to August 6; 40 records, April 21 to May 25, indicating the height of the season." In a multi-brooded species the "height of the season" is actually the first brood. Later records must pertain to second or more broods. Furthermore, it is difficult to believe that a given population subjected to the same environmental factors would respond with such irregularity as is indicated by egg dates from 10 March to 21 April. These extreme dates must come from different populations or from first and second broods.

As we pass on to larger areas in Arizona, the confusion increases. If we attempt to correlate nesting with temperature, then elevation must be taken into account, for the range of the Cactus Wren extends from near sea level at Yuma to the lower edge of the Upper Sonoran Life Zone in the eastern part of the state. Are eggs laid earlier at lower elevations where the temperatures rise earlier in the season? Not necessarily, for the rainfall in Arizona is directly proportional to elevation; and rainfall is necessary to produce the spring annuals and abundant insects. Hensley (MS) gives the date of 10 March 1949 for the first egg of the Cactus Wren in Organ Pipe Cactus National Monument in southwestern Arizona. The area which he studied is slightly lower in elevation than Tucson, but the mean temperatures for the first 3 months in the year are slightly higher. The annual precipitation is about 3 inches (76 mm) less. Unfortunately, we do not have any nesting from Tucson for 1949 for comparison.

Through the courtesy of the late Fred M. Dille, we have available the field note book of George F. Breninger, who collected eggs in the Phoenix region of Arizona from 1896 to 1905. The elevation of Phoenix is about 1,200 feet (360 m); the mean temperatures are higher than those in Tucson, and the average annual precipitation is similar to that in parts of Organ Pipe Cactus National Monument. In five of the years data are fairly complete, although the condition of the eggs was not always recorded. The dates on which the first complete sets of eggs were collected are as follows: 10 March 1896, 8 March 1897, 11 March 1898, 21 March 1899, and 21 February 1901. Most of the subsequent sets collected were in the month of March, and some of these later sets were catalogued as advanced in incubation. The habitat is not described, but nests were found in mesquites, "thorn bush," ironwood, palo verde, and "thorn tree"; a few were in cacti. These dates are, of course, subject to the same errors as those in Bent's "Life Histories." They are not dates of first eggs laid, but dates when eggs were collected. It is to be noted that none of these dates are in April.

In contrast to these old records, we have a letter from Ruth M. Crockett, of Phoenix, reporting that in 1952 nestlings were being fed on 3 February, near Squaw Peak. The young in another nest were fledged about 23 February. We estimate that the first of these eggs was laid about 13 January. The nest locations were described as on the edge of a residential area, and "the vegetation was unusually lush from recent rains."

There has been little in the foregoing extensive summaries to support the supposition that the time of egg laying is genetically controlled for a given population. Some evidence, we believe, has been presented to show that most of the records in Arizona occur in the month of March, regardless of elevation, and that earlier egg laying takes place under favorable circumstances. This was the status of our studies up to the beginning of 1953, when we decided to turn our attention to a larger population of Cactus Wrens on the Santa Rita Experimental Range, a desert tract situated about 35 miles (56 km) south of Tucson. There we discovered that egg laying began about a month later than in the Tucson region. To our great surprise, not one pair of Cactus Wrens laid their first eggs before 16 April. This held true also in 1954, 1955, and 1956. The Curve-billed Thrasher, likewise, was late in laying. Nice (1937: 104) found that the nesting of the Song Sparrow did "not depend on the state of the vegetation." Our two places of study were not identical and comparisons are difficult. The ubiquitous annual bladder-pod which often, very early in the spring, carpeted considerable portions of the nesting territory at Tucson, was absent on the Range. The creosote bush, a major part of the environment at Tucson, was also missing. In both places nest building began before the larger perennials were in bloom (Table 8.2).

Table 8.2. Santa Rita Experimental Range.
Date of first Cactus Wren egg laid (estimated).

Year	Number of nests	Average date first egg	Range	Standard deviation
1953	16	29-4	17-4—13-5	7.35
1954	5	2-5	24-4—21-5	9.65
1955	5	28-4	16-4—11-5	9.78
1956	8	3-5	21-4—13-5	7.41

The increase in elevation from our Kleindale Road home at Tucson to the upper edge of the Pasture 5 research plot is approximately 900 feet (270 m). Assuming a thermal lapse rate of 1F (.56C) for each 250 feet (75 m) of elevation (Lowe, 1964: 85), the Pasture 5 area should be 3.6F (2C) cooler than the Tucson locality. We checked the temperatures for the months of January, February, and March at a recently installed thermograph, just east of Pasture 5, at an elevation of 3,350 feet (1005 m). They show the Range daytime temperature to be 3F (1.7C) lower. Curiously, a temperature inversion, brought about by the different topography, causes the night temperatures to be 3F (1.7C) warmer. Thus the daily mean temperatures of the two stations are identical. Obviously, if we endeavor to base the time of egg laying upon a period of rising mean temperatures as we did at Tucson (Table 8.1), the dates would coincide. The monthly mean temperatures during the period of our study, from a thermograph station situated about 3 miles (4.8 km) south and 400 feet (120 m) higher proved to be too variable to permit any safe generalization. Nevertheless, it is difficult to escape the feeling that temperature somehow is a factor in the spring nesting.

In 1958, after a mild winter, egg laying occurred unusually early at Tucson. On 4 May 1958, we made a hurried visit to our old area on the Range to check the situation again. We found two nests with young. At one of them, three of the occupants popped out and flew to safety. We captured the fourth; its measurements indicated that the date of laying of the first egg must have been about 29 March. On an average, in the spring, the arrival of a given maximum temperature on the Range would lag behind that at Tucson by some days. Furthermore, the daily variation of temperature is 6F (3.3C) less on the Range. The Cactus Wrens are not only subjected to a lower daily rate of change of temperature, but also to a time delay in maximum temperatures. We cannot prove that these two conditions are of sufficient importance to produce a lag of a month in egg laying. The effect of environmental temperatures upon the physiology and behavior of the Cactus Wren is still a relatively unexplored field. There may be a distinct population on the Santa Rita Experimental Range whose time of egg laying is genetically controlled to occur a month later than that on the Tucson mesa.

In the Saguaro National Monument study area variations in the date of the first egg of the first brood (Table 8.3) were indeed frequent, and were obviously not the exception. The range varied from a low of 7 days in 1964 to a high of 47 days in 1965; the average was 28.1 days. The standard deviation is low in the years when the population was low. It increases at higher densities, but the number of examples seem too few to permit any generalization. Competition by year-old wrens for space in a previously stable situation could probably cause delays in nesting. In 1965 the first eggs were laid later in six of the eight territories established by year-old males than that year's average of 27 March; the adults

Table 8.3. Saguaro National Monument. Date of first Cactus Wren egg laid.

Year	Number of nests	Average date first egg	Range	Standard deviation
1963	8	30-3	17-3— 8-4	5.93
1964	10	18-4	15-4—22-4	2.49
1965	18	27-3	11-3—27-4	12.09
1966	9	24-3	17-3— 4-4	7.83
1967	15	5-5	15-4—20-5	10.64
1968	14	13-3	24-2— 4-4	13.56

laid earlier. Because of fewer known year-old wrens this tendency is less evident in 1967 and 1968.

Five females, whose dates of hatching are known, laid their first egg when less than a year old, as follows: CF-66, 6 May 1964, 27 April 1965, age 356 days; CF-121, 12 April 1965, 31 March 1966, age 353 days; CF-230, 27 June 1966, 18 May 1967, age 325 days; CF-234, 26 June 1966, 19 May 1967, age 327 days; and CF-265, 14 May 1967, 6 March 1968, age 280 days.

The climatic factors which influenced the time when egg-laying began appear to be, as in the earlier Tucson study, a combination of temperature and rainfall (Table 8.4). Precipitation varies considerably over the Tucson basin. We have used the values recorded at the nearest gage, at Saguaro National Monument headquarters located about 2

Table 8.4. Saguaro National Monument. Climatic factors influencing start of Cactus Wren egg laying.

Winter	Average date first egg	Rainfall Oct-Feb	Rainfall Jan-Feb	Temperature departure Jan	Feb	Av. of mean temp. of 14 days preceding mean laying	Growth annual plants
1962-63	30-3	5.00 (127)	3.00 (76.2)	−1.3 (−.7)	+4.4 (+2.4)	61.2 (16.2)	Good
1963-64	18-4	2.95 (74.9)	.81 (20.6)	−2.1 (−1.2)	−4.4 (−2.4)	63.8 (17.7)	Poor
1964-65	27-3	5.06 (128.5)	1.63 (41.4)	+3.3 (+1.8)	−.8 (−.4)	58.3 (14.6)	Good
1965-66	24-3	12.38 (314.5)	4.39 (111.5)	−1.3 (−.7)	−4.1 (−2.3)	62.8 (17.1)	Good
1966-67	5-5	1.13 (28.7)	.21 (5.3)	−.8 (−.4)	+1.5 (+.8)	61.4 (16.3)	Very Poor
1967-68	13-3	8.57 (217.7)	3.19 (81.0)	+4.2 (+2.3)	+7.8 (+4.3)	60.6 (15.9)	Good

Rainfall in parentheses is in millimeters; temperature in parentheses is in degrees Centigrade.

miles (3.2 km) south of the research area, at an elevation of 3,100 feet (944.9 m). Since spot checks of the temperatures in the research plot indicate that the mean temperatures parallel those at the University of Arizona station in Tucson, averaging slightly lower, 2 to 3F (1.1-1.7C), with lower maximums and minimums and a similar range (Green and Sellers, 1964), we believe the University values can be used here without serious error.

Laying always followed a warm "wave," a steep rise of 10 to 20F (5.6-11C) in the mean temperatures, of from 10 to 14 days' duration. In the 4 years with sufficient winter rainfall to produce an abundant growth of annual plants, the average date of laying occurred in March. Low rainfall at the beginning of 1964, resulting in a poor crop of annuals, coupled with below-normal spring temperatures, apparently delayed laying until April that year. 1967 was worse. The wrens laid in May, after a winter of extremely deficient rainfall, a very sparse growth of spring annuals, and below-normal temperatures in March and April.

Some incomplete records are available for 1962. (Regular visits did not begin until August 1962.) The January-February rainfall totaled 1.97 inches (50 mm); temperature departures, above normal, were 1.3F (.72C) for January, 3.6F (2.0C) for February. From 2 to 11 February the mean temperatures rose from 56 to 69F (13.3-20.6C). Dates of first egg laid are 15, 18, and 20 February and 1 March in territories III, south of VIII (extralimital), VI, and V respectively. These dates represent the earliest layings in the 7 consecutive years.

Comparisons with the suburban Tucson egg data (Table 8.1) are not entirely satisfactory, for, despite the long record, none of the years coincide with those worked in the Saguaro National Monument. Eighteen layings occurred in March before the 27th, six in February, and one even in early January. The average date, omitting the obviously abnormal 2 January 1958, is 8 March. Four additional first egg records in the single territory on Kleindale Road, 12 February 1962, 13 March 1964, 25 February 1965, and 22 February 1967, are all earlier than the Monument average, earlier even than the earliest in the Range. The average date of the first egg in the Monument, that is, the average of the annual averages, is 4 April, a rather meaningless figure since only six samples are available, and the yearly variations are large. The difference in elevation between the two localities is less than 500 feet (152.4 m); the temperatures, slightly different, vary in parallel; and precipitation, although variable, usually occurs on the same days. At present the evidence again suggests that layings occur earlier in and near cities, and that favorable warm temperatures, inducing an increased insect supply from the extensive well-watered ornamental plantings, is the cause. In a typical desert habitat, such as Saguaro National Monument, the insect population is dependent upon growth of the winter-annual vegetation, which in turn varies with rainfall and temperature. Nesting begins usually before the perennial plants flower.

9. The Eggs of the Cactus Wren

The eggs in a Cactus Wren's nest cannot be seen from the outside, even with a mirror, for they are sometimes virtually hidden among the feathers on the floor of the nest cavity. We made no systematic attempts to remove them for inspection, weighing or marking, for the possibility of breakage during the handling in the confined nest space was too great. Our limited population of Cactus Wrens offered little opportunity for comparative studies of size, color, or shape of eggs. Bent (1948: 225) describes the eggs as "mostly ovate in shape, some being slightly elongated or shortened. They are somewhat glossy. The ground color varies from 'salmon color' or 'salmon-buff' to 'seashell pink,' pinkish white, or rarely to nearly pure white. Usually the egg is more or less evenly covered with fine dots or very small spots of reddish browns, 'rufous' to 'ferruginous,' sometimes nearly concealing the ground color; sometimes the markings are concentrated in a ring about the larger end. Rarely an egg seems nearly immaculate, and still more rarely an egg with a white ground is quite heavily spotted or blotched with the above browns. The measurements of 50 eggs in the United States National Museum average 23.6 by 17.0 millimeters; the eggs showing the four extremes measure 26.4 by 15.2, 24.9 by 19.1, and 19.8 by 13.2 millimeters."

We counted the eggs by inserting a hand into the vestibule and enlarging it by expanding the fist and fingers as the hand moved inward and downward. The spiny, easily detached joints of the jumping cholla were especially vicious and treacherous. Pliers with which to remove the spines from hands, wrists, and forearms were always carried along. Finally an aluminum sleeve was constructed which afforded considerable protection. It was thrust into the nest entrance, but it was still necessary to use the fingers to enlarge the diameter. Frequently some of the joints of the cholla had to be bent aside or removed to facilitate inspection. What was an easy, accessible route for the occupants was often a dangerous and difficult one for larger intruders. The damage to the nest and its surroundings during inspection was sometimes rather extensive. Nevertheless, desertions did not occur, even when the visit took place just before the first egg was laid or immediately after.

The female began roosting at night in her newly constructed breeding nest as soon as it was well covered over and some of the lining was in place. Sometimes she occupied it a week before the first egg was laid. Meanwhile, the installation of additional lining continued, but at a reduced pace, by both adults. In those cases where the female appropriated the male's roosting nest for her breeding nest, the refurbishing of the interior began anew. For example, in February of 1959 it was not entirely clear whether the nest a female took over was at first intended to be a roosting nest. It had been chosen because of the many disturbances the female suffered in her winter quarters in the pyracantha bush near our front porch with its electric light. She occupied the male's roosting nest for 11 nights before she laid the first egg of the year.

Eggs were laid at the rate of one a day on consecutive days in early morning after sunrise. Possible exceptions to this routine occurred in 1941. Two eggs in nest 35B were found on 25 February at 1745. The third egg was laid on the 26th after 0720. On the following day the nest was visited at 0715 and 1745, but only three eggs were found. The female was frightened out at the first visit, and may have had to lay the egg before returning to the nest. We found the fourth egg on 28 February. The fourth clutch of 1941 contained two eggs on the evening of 12 June. On the 13th at 1900, we found a broken egg on the ground beneath the nest, and there was only one egg in the nest. No further laying occurred at least up to 1850 on the 14th. When we next visited this nest at 1900 on the 15th, another egg had been laid.

Unfortunately we did not find the first egg of 1959 until late afternoon. The second egg was laid on 20 February between 0730 and 0830; the third was laid on the 21st between 0748 and 0850. In March of 1947 the fourth and last egg of the clutch was laid between 0723 and 0801, the others in order, before 0725, 0730, and 0720. We doubt that a female remains very long in the nest after she has laid. In 1947, the time required to lay the fourth egg after the female entered her nest was 38 minutes; in 1959, it was 62 minutes for the third egg. These would be extremely long intervals, especially since they occurred early in the morning when the wren needed food. Therefore the layings in 1947 and 1959 probably took place just before the female was seen to leave the nest. In March of 1944, two eggs of one clutch were laid before 0720 on consecutive days; the next egg was laid after 0720. A similar situation, just noted, was observed in 1947, when the fourth egg was laid somewhat later in the morning than the first three. The first eggs of two clutches in June and July, when day length is greater, were laid earlier than those in February and March, but the last eggs of these clutches were held until after 0800 and 0710, respectively. Apparently as laying progresses, the time required for egg formation increases to more than 24 hours.

Available data on the number of eggs in each annual series of clutches, which we found in nests in the Kleindale Road area, are shown in Table 9.1. When nests were inaccessible, as happened occasionally in neighbors' lots, we were compelled to estimate the number of eggs later, from the number of fledglings we saw being fed by the parents. Thus these estimates are minima, for the actual number of eggs may have been larger. An assumption that a minimum of three eggs was laid seems safe. We have never found less than three eggs in a complete set. In 13 of the 22 years, the first clutches contained three eggs each; in the other 9 years, there were four eggs in each first clutch. The average is 3.41 eggs per clutch. Bent (1948: 225) reported that the usual set "consists of four or five eggs, most commonly four; but as few as three may constitute a full set, and as many as six or even seven have been found in a nest," but no tabulation of dates, or which brood was involved, is given. Brandt (1951: 187) says "the usual clutch numbers three or four, with an

Table 9.1. Kleindale Road area. Number of eggs per clutch of Cactus Wren.

Year	1st	2nd	3rd	4th	5th	6th
1939	3(est.)	3	3(est.)	4		
1940	4	4				
1941	4	5	5	3	4	
1942	4	3(est.)				
1942	4	5				
1943	3					
1944	3	1	4	3(est.)		
1944	3					
1945	3	4	4			
1945	4	4				
1945	3					
1947	4	3(est.)				
1947	4	3(est.)				
1948	3(est.)					
1952	3(est.)					
1953	3	3	3			
1954	4	3				
1955	3(est.)	3(est.)	3(est.)			
1956	4	4	4	4		
1957	3	4	4			
1958	3	4	5	5	4	4
1959	3	3	3			
Total Eggs	75	58	38	19	8	4
Average Eggs per Clutch	3.41	3.63	3.80	3.80	4.00	4.00

The 1944 clutch with one egg has been omitted in the totals and averages.

infrequent five." For comparison we summarize a total of 24 sets of eggs collected in the months of February and March by George F. Breninger (field note book) near Phoenix, Arizona, in 1896, 1897, 1898, 1899, and 1901. There were 10 clutches with 3 eggs each, 12 with 4 eggs each, and 2 with 5 eggs each. The average was 3.67 eggs per clutch.

Brandt (1951: 190) has suggested that in a period of prolonged drought, such as preceded the laying in 1949, the low number of three eggs, which he found in a nest near Rillito Creek, may have been due to the wren anticipating a reduced food supply for its offspring. There is little in our data to support this view. Precipitation from 1 October 1939 to the laying of the first egg on 20 February 1940 was 2.08 inches (52.8 mm). The following winter almost four times as much rain was recorded in a similar period. Yet the clutches for each of these years contained four eggs each, the first eggs being laid on 20 and 24 February, respectively. In the spring of 1953, after 4.84 inches (122.9 mm) of rain, a set of three eggs was laid. Then, a month earlier in 1954, and after but 1.21 inches (30.7 mm) of rain, a clutch of four eggs was laid.

In Table 9.2 we have rearranged the 1939 to 1959 Kleindale Road figures of Table 9.1 to show the relative number of clutch sizes. The second clutch of 1944, containing only one egg, has been omitted. There were no clutches of two eggs. With four exceptions, these figures pertain to one territory per year.

Our week-end visits to the Santa Rita Experimental Range did not, unfortunately, enable us to obtain exact data on the number of eggs laid in each nest. We estimate that three or four eggs was the usual number. The late spring laying also made second and third broods uncertain.

In Saguaro National Monument in the course of the 6 years, Cactus Wrens laid 154 clutches of eggs (Table 9.3), each pair averaging two clutches per territory per year. Included in this total are replacements of destroyed clutches, and also several extralimital clutches, the latter because the wrens moved slightly beyond the boundaries to lay their second or third clutches, while they maintained their original territories. Territories XX, 1965 (male only), XXI, 1967 (banded pair present briefly at beginning of season), and XV, 1968 (female disappeared) are omitted because no layings occurred. In the total of 78 territorial seasons, 19

Table 9.2. Kleindale Road, 1939 to 1959.
Number of Cactus Wren clutch sizes.

Figures in parenthesis are estimated sets (3 eggs each) included in totals.

Number of nests with					Average
3,	4,	5 eggs	Nests	Eggs	set
27 (11)	24	5	56	202 (33)	3.6
Percentages					
48.2	42.9	8.9			

Table 9.3. Saguaro National Monument. Cactus Wren clutches per territory.

Territory	1963	1964	1965	1966	1967	1968	Total
I	2	2	2	3	1	2	12
II	2	2	2		2	3	11
III	1	1	2		1	3	8
IV	2	2	2	3	2	3	14
V	2	2	2	1	1	1	9
VI	2	2	2	3	2	3	14
VII	3	2	2		2	3	12
VIII	2	2	2	2	2	3	13
IX	1	2	2	3	2	2	12
X		2	2		1	3	8
XI		1	2		1	1	5
XII			1				1
XIII			2	3			5
XIV			2		1		3
XV			2	3			5
XVI			1	2			3
XVII			2		1	3	6
XVIII			2				2
XIX			1		1		2
XX							0
XXI						3	3
XXII					1		1
XXIII						2	2
XXIV						3	3
Total	17	20	35	23	21	38	154
Average per territory	1.9	1.8	1.8	2.6	1.4	2.5	2.0

wrens laid only one clutch (24.4 percent), 42 wrens laid two clutches (53.8 percent), and 17 wrens laid three clutches (21.8 percent) in a single season. Table 9.4 lists the percentage of single, double, and triple clutches in each year. Territories I to XI contained 76.6 percent of all the clutches laid, the remaining 13 territories evidently being occupied only in years of good survival of surplus populations. (The extralimital territory south of XIII had four clutches in 1968.)

Table 9.4. Saguaro National Monument. Percentage of single, double, and triple Cactus Wren clutches.

Clutch	1963	1964	1965	1966	1967	1968
Single	22.2	18.2	15.8	11.1	60.0	13.3
Double	66.7	81.8	84.2	22.2	40.0	20.0
Triple	11.1			66.7		66.7
	100.0	100.0	100.0	100.0	100.0	100.0

The late nestings in 1964 and 1967 reduced the average clutches per territory 10 percent in 1964 and 30 percent in 1967 below the six year average (Table 9.3). In 1967, 60 percent of the territories had single clutches. Apparently time was a factor in 1967, for the average date of the first egg laid was very late. With early layings in the 15 territories of 1968, the number of clutches and the average clutch per territory almost doubled over those of the preceding year. The near average figure of 1965 (the same as in 1964), and the extreme high of 1966, are more difficult to explain, because the wrens laid early in both years and at approximately the same time.

There is very little in the averages of Table 9.3 to support the view held by Kendeigh (1934: 309), that the number of clutches per female per season varies inversely with the number of breeding females. Kendeigh, faced with an exception in his data on the abundance and reproduction of the House Wren (*Troglodytes aedon*), assumed a time-lag in reproduction following a drastic decline in the population. The Cactus Wrens showed a 5.3 percent decrease in average clutch per territory when the population increased from 9 to 11 territories, from 1963 to 1964. However, an increase in 1965 to 19 territories produced no change in the average number of clutches. Then, in the decline in 1966, we find a startling increase of 44.4 percent in the average, the value reaching 2.6 compared to 1.9 in 1963. The inverse correlation held in 1967, when the increase to 15 territories produced a drop of 46.2 percent in the average number of clutches, but it failed in the succeeding year when the same number of territories produced a rise of 78.6 percent. The probability is, that had nesting not been delayed in 1964 and 1967, the average number of clutches per female would have been larger and the correlation less inverse.

In 1965 the wrens in the 16 territories laid the first egg of their second clutches between 1 and 28 May; the average date was 13 May. (A replacement of a lost male in territory VIII after the first nest was robbed did not delay the laying of a second clutch.) After the June-July fledgings the wrens did not attempt a third brood. Events in the three single clutch territories went as follows: the year-old male in territory XII disappeared a month after the fledging of the first brood. A very late start, 4 May, in territory XIX, ended in failure when the nest was robbed. The year-old male vanished a month later. Unfortunately the females in both these territories were nobands; they could not be traced, and no replacements appeared to occupy the vacant territories. The banded year-old pair in territory XVI laid late in April and raised only one brood.

In 1966 the banded male and his noband mate in territory IV disappeared when their nest was robbed. The banded pair in adjacent territory V, whose first nest had suffered loss, then moved into IV to lay their second clutch. The average date of the second clutches in the

Table 9.5. Saguaro National Monument. Number of Cactus Wren clutch sizes.
Figures in parentheses are estimated sets (3 eggs each) included in totals.
Extralimital nests are excluded.

Year	2,	Number of nests with 3,	4,	5 eggs	Nests	Eggs	Av. set
1963	1	11(5)	5		17	55	3.2
1964	1	17(8)	2		20	61	3.1
1965		24(6)	11		35	116	3.3
1966		11(9)	10	2	23	83	3.6
1967	5	12(3)	4		21	62	3.0
1968	2	14(2)	19	3	38	137	3.6
Total	9	89(33)	51	5	154	514(99)	3.3
Percentages							
1963	5.9	64.7	29.4				
1964	5.0	85.0	10.0				
1965		68.6	31.4				
1966		47.8	43.5	8.7			
1967	23.8	57.1	19.1				
1968	5.3	36.8	50.0	7.9			
Total	5.8	57.8	33.1	3.3			

remaining eight active territories was 8 May. The relatively late nesters in territories VIII and XVI (17 and 18 May, respectively) laid only two clutches; the other six pairs laid three clutches, most of them by the middle of June.

The Cactus Wrens laid a grand total of 514 eggs in their 154 clutches, averaging 3.3 eggs per clutch (Table 9.5). This total includes 33 nests, most of which were inaccessible, estimated to have had three eggs each. The error introduced in subsequent calculations varies with the number estimated, but it probably is on the conservative side. A few nests may have had two eggs; it is more likely that some had four.

Six females, who laid double clutches, laid three eggs in their first clutch and four in the second; two laid four eggs in their first and four in their second; nine made no change; three laid four eggs in their first clutch and three in the second. The inclusion of estimated egg totals leaves the disposition of the remaining layings uncertain. Eleven of the 17 triple clutches (two in 1966 and nine in 1968) did not require estimates. Their eggs total 39 in the first clutch, 46 in the second, and 39 in the third. If the remaining six triples, with their eight estimated clutches, are included, the totals rise to 60, 65, and 58. The 11 triples reveal a tendency toward more eggs in the second clutch than in the first, or the same number; in one triple there was a drop in the second set, as follows: 4-5-4, 3-5-3, 3-5-4, 3-4-3, 2-4-3, 5-5-4, 4-4-3, 4-4-4, 4-4-4, 3-2-3.

The average number of eggs laid per female was 6.1 in 1963, 5.5 in 1964, 6.1 in 1965, 9.2 in 1966, 4.1 in 1967, and 9.1 in 1968. The maximum, 14 eggs by CF-174, occurred in 1966. Females laying only one clutch averaged 3.1 eggs per clutch; those laying two clutches averaged 6.5 eggs, and those laying three clutches averaged 10.8 eggs. Single clutches contained a total of 59 eggs, doubles, 272 eggs, and triples, 183 eggs.

In the course of the first 3 years in the Saguaro National Monument research area, egg production rose as the number of breeding pairs increased. By increasing the average eggs per clutch and the average clutch per female, the egg production of the nine breeding pairs in 1966 exceeded that of the same number of pairs in 1963. A similar situation prevailed from 1967 to 1968. The egg production and the average eggs per set of the 11 pairs in 1964 and 15 pairs in 1967, after late starts, were almost identical, but the average clutches per female declined in 1967, thus balancing the production of the 2 years.

Table 9.6 gives the composition of the first, second, and third clutches laid in the Monument research area. There was a 22.1 percent drop in the number of breeding nests from the first to the second clutch, and 71.6 percent from the second to the third. The average eggs per clutch increased slightly as the season advanced. A greater increase, 3.41 to 3.63 to 3.80, occurred in the Tucson area (Table 9.1), with a somewhat larger drop, 27.3 percent, in the breeding nests from the first to the second clutch, but a much smaller one, 37.5 percent, to the third. Evidently egg production was considerably better in this locality. Comparisons, however, are of dubious value, for the two studies were conducted at different times, and none of the years coincide.

Table 9.6. Saguaro National Monument. Composition of first, second, and third clutches of Cactus Wren.

Figures in parentheses are estimated nests (3 eggs each) included in totals.

Clutch	2,	Number of nests with 3,	4,	5 eggs	Total	Av. eggs per clutch
1st	6	44(4)	26	1	77	3.29
2nd	3	35(24)	18	4	60	3.38
3rd	0	10(5)	7	0	17	3.41
Total	9	89(33)	51	5	154	
Percentages						
1st	7.8	57.1	33.8	1.3	50.0	
2nd	5.0	58.3	30.0	6.7	38.9	
3rd		58.8	41.2		11.1	

10.   Incubation

Incubation was performed entirely by the female. So far as we could observe, the male never entered and remained in the breeding nest when the female left to search for food. The nest was occupied by the female the night after the first egg was laid, and the female continued to occupy the nest nightly without interruption thereafter. Daytime incubation was extremely irregular on the first day, after the first egg was laid. In fact, it is doubtful if effective warming occurred at all. In nests which we inspected on the first day in mid-morning, at noon, and in the afternoon, we found the egg decidedly cold to the touch and the female was absent. Cold eggs were also found in some nests in the early forenoon of the second and third day. Does embryonic development begin as soon as the first egg is laid, or does it await the initiation of regular, attentive incubation after the final egg? The Cactus Wren could sleep in the adequately long vestibule of her nest, awaiting completion of egg laying, without transferring any of her body heat to the egg, but we do not believe she does this. We seldom inspected breeding nests at night, for a frightened wren would have great difficulty in finding its way back. However, a wren which was accidentally disturbed at 2055 on 27 April 1941 left a very warm, single egg in the nest. Evidently she had been sitting on the egg. Swanberg (1950) has shown that some species of birds can actually sit in the nest without warming the eggs appreciably. In a careful study of the European Blackbird (*Turdus merula*), Enemar (1958) found that incubation began before the clutch was complete. When the Cactus Wren enters her breeding nest, she retreats to the far end into the depressed cavity, just as she does in her roosting nest. It would seem illogical to assume that a bird which for the greater part of the year regularly roosts in a deep cavity would, when the first one or two eggs are laid, foresake this comfortable place, to roost in the narrow, tunnel-like vestibule in which it cannot turn around. As further evidence of early incubation and warming, we have the observed fact that all of the eggs never hatched on the same day.

Using Heinroth's (1922) rule and its elaboration by Swanberg (1950) for the determination of the incubation period, we checked the time from the laying of the last egg to the hatching of the last young. Most nests were inspected twice a day, in the early morning and in the early evening. Some could be visited only in the late afternoon. The period of incubation was 16 days in nine nests for which we have accurate data. In three other nests with incomplete data, we estimate it to be also 16 days, but there is a possibility it could have been 15.

Our observations are at variance with those of Hensley (1959: 89), who reported that "incubation lasted 17 days in two nests." In one of these, however, he states that "the clutch of three eggs was completed on March 29 and on April 16 the nest contained three newly hatched young." If the last one hatched on 16 April, the incubation period was actually 18 days.

We were able to study incubation attentiveness on Kleindale Road in considerable detail at nest 19C in 1947. Data from 2 complete days, from the awakening of the female to her retirement, are available. In addition, we have another day with only 18 minutes missing. We found the first egg on 26 March when we examined the nest at 0725. Both adults carried lining material to the nest from 0925 to 1015. On the last trip the female, HF-50, remained in the nest 5 minutes. On 27 March we watched the nest from 0650 to 0937. The six periods on the nest varied in minutes as follows: 10, 15, 7, 6, 0.5, 5. The periods off the nest were: 15, 16, 7, 5, 53.5, 27. Average time on the nest was 7.3 minutes; average time off was 20.6 minutes. We checked the nest at 0730; it contained two eggs. The female flew out as we approached. Evidently incubation was already in progress, but it was very irregular, with only 26 percent of the time devoted to warming the eggs. HF-50 carried bits of lining material to her nest at every visit. At long intervals the male brought food to her as she sat inside. We found the third egg on 28 March at 0720, but no further observations were made that day.

On 29 March we began our watch before sunrise. Our observation post was a bedroom window facing the nest only 15 feet (4.5 m) away. It was more comfortable than a blind; the wrens suffered no disturbance, but it had the disadvantage of narrowing our field of view, so that the wrens could not be watched after they left the nest. We remained at our post continuously until 1722. After a short absence, we began again at 1740 and watched until HF-50 retired at 1848. The first song of the male came at 0558. HF-50 left her nest 19C at 0603. The sky was clear and there was no wind. The official University of Arizona minimum temperature was 46F (7.8C). We inspected the nest at 0615, when the female was absent, and counted three eggs. Eighty minutes elapsed before HF-50 returned to her nest. Meanwhile her mate had visited it once and entered, but he left and continued singing in the vicinity. The female now remained in the nest for 38 minutes. The fourth and last egg was probably laid at this time. (We did not inspect the nest until 1450; it contained four eggs then.) By 1722 HF-50 had sat on the eggs 21 times and had been off the nest 22 times. The average time on the eggs was 11.6 minutes; average time off was 19.8 minutes. The length of the periods on the nest varied from 1 to 42 minutes; off periods varied in length from 5 to 80 minutes. HF-50 incubated 35.9 percent of the daylight time. Following our resumption of observations at 1740, there was one more period on the nest of 9 minutes, beginning at 1827, and one off period of 12 minutes, after which the female retired. The addition of these two periods to the total has only a slight effect on the averages just presented. On the succeeding 6 days lack of time limited our observations to about an hour each day in the forenoon. The data we obtained are too fragmentary to permit any safe conclusions, except that the lengths of the extreme on-and-off periods appeared to decrease.

We watched nest 19C for a whole day on 5 April, the seventh day of incubation of the full clutch, and recorded 28 attentive periods and 29 inattentive periods from 0617 to 1850. The former averaged 14.8 minutes; the latter averaged 11.7 minutes. The range of variation was 1 to 28 minutes and 2 to 26.5 minutes, respectively. For the distribution of these on-and-off periods in the course of the day, see Fig. 10.1. In the wave shape of the "on" periods, the periods occur in groups; a group of long periods is followed by a group of short ones, and these are followed by another group of long ones, the peaks and depressions decreasing and smoothing out in the afternoon. The cyclic behavior is present, but not so pronounced in the curve of the "off" periods. There is a distinct lengthening trend in the "off" periods as the day becomes warmer. The temperatures on 5 April were: maximum 70F (21.1C), minimum 38F (3.3C), and mean 54F (12.2C). The length of HF-50's day of activity was 12 hours and 33 minutes; 54.9 percent of this time was spent in incubation.

On 12 April 1947, the day before the first egg hatched, the female left nest 19C at 0556. She retired at 1852. In the course of this day of 12 hours and 56 minutes, she devoted 50.3 percent of her time to the nest. There were 27 attentive periods averaging 14.5 minutes each, and 28 inattentive periods averaging 13.8 minutes. The variation was from 1 to 54 minutes in the attentive periods, and 6 to 26 minutes in the inattentive ones. Fig. 10.1 shows the distribution of these periods. Temperatures on this day were: maximum 79F (26.1C), minimum 48F (8.9C), and mean 63.5F (17.5C). With the exception of the first part of the day, the curve of "on" periods has smoothed out considerably. The extremely long attentive period of 54 minutes began at 1000. It is difficult to explain, but it may have been induced by an immature Curve-billed Thrasher who climbed into the cholla and moved about beneath

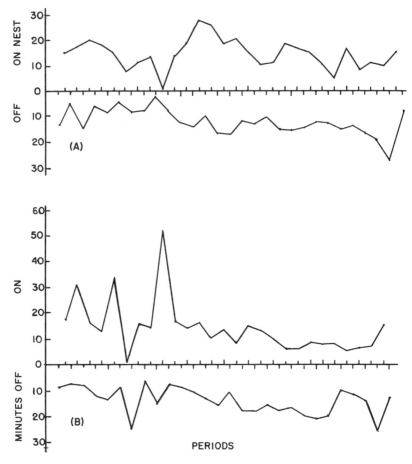

Fig. 10.1.   Incubation attentiveness at Nest 19C in Lot 7 on Kleindale Road, Tucson, Arizona. A, 5 April, 1947; B, 12 April, 1947.

the wren's nest at this time. We have observed that Cactus Wrens will enter and remain inside their breeding nests when there is a possibility of intrusion by neighboring birds.

Far more difficult to account for is the occurrence of the 1 minute attentive periods. There was one each on 29 March, 5 April, and 12 April. No disturbances of any nature could be observed at these times. On 12 April, the male, HM-54, visited nest 19C eight times between 0707 and 1350. Half of these visits occurred when the female, HF-50, was in the nest; the male then remained in the vestibule from 0.5 to 1.5 minutes. On those visits when he discovered that the female was absent, he left at once. On his last trip at 1357 he carried a small object, apparently an insect, into the nest, but HF-50 was not there. Evidently he swallowed the insect, for he left without it. At 1841 the four eggs had not yet hatched.

Armstrong (1955: 174-184) gives an extensive summary of the incubation of the European Wren, *Troglodytes troglodytes,* with comments on the researches of Allen and Nice (1952), Baldwin and Kendeigh (1932), Heath (1920), Laskey (1946), and Nice and Thomas (1948) on several of the smaller American wrens. The inverse relationship between the attentive periods and the temperature seems to be the general rule. The average body temperature of eight species of passerine birds is reported by Baldwin and Kendeigh (1932: 157) to be 107.6F (42C) during the daylight incubation period. Referring to the House Wren, Nice and Thomas (1948: 150) say, "its eggs are best developed at a temperature of 100F (37.8C) or below, rather than above, and that a lower temperature is required during the early part of the incubation in order to begin the development, than is desirable to continue the development through the later stages." None of the above investigations were concerned with the extremely high temperatures encountered in the summer nesting season in the North American deserts. At Tucson, in the latter part of June and the first part of July, before the rains appear, maximum temperatures, day after day, usually range from 105 to 110F (40.5-43.3C) or even more. The Cactus Wren's nest cavity can often reach these ambient temperatures, or even higher ones occasionally, as attested by our Weston thermometer. At such times, incubation must obviously become a cooling procedure, not a period of warming.

Baldwin and Kendeigh (1932: 139-141), studying the resistance of embryos of the House Wren to high temperatures, concluded that "there is a 50 percent likelihood that egg embryos partly incubated will survive an hour's subjection to air temperatures of 106F (41.1C) or 111F (43.9C), but that they will all be killed at 114F (45.6C) or 116F (46.7C)."

We confess to a deplorable lack of observations on incubation attentiveness during the hottest part of the summer days. We conducted all of our Kleindale Road and Santa Rita Experimental Range studies on our spare time. In most of the years our 2- to 4-week out-of-state vacation trips in June or July unfortunately left a gap in our records. Furthermore, after the senior author once experienced the acute discomfort and fright of heat exhaustion, he was in no mood to risk a repetition in a remote area. Summer visits to Saguaro National Monument usually ended before noon.

We do have a few observations of wrens entering a nest to incubate in mid-afternoon on very hot days. On 9 July 1958 at 1900, when the temperature was 104F (40.0C), we found a female wren at the entrance of her breeding nest. The incubation rhythm may be partly determined by hunger, for a bird must leave its nest to find food. We doubt that a Cactus Wren is capable of estimating the proper temperature of her eggs.

HF-50 continued to bring bits of lining material, such as fine grasses and small feathers, to her nest throughout the entire incubation period. For the first week, the male assisted in this work, but he did so very irregularly. He visited the nest occasionally, without nest material, apparently out of curiosity or in search of his mate. When she was absent, he would enter, but he remained inside only a few seconds before flying away. Courtship feeding was seldom observed. It probably occurred no oftener than three or four times a day at the beginning of incubation, and apparently it ceased altogether before the eggs hatched. We did not note any feeding on 12 April.

The time required for all the eggs to hatch in the completely successful nests varied as follows: in three nests with three eggs each, the time was 3, 2, and 2 days; in five nests with four eggs each, the time was 2, 2, 3, 2, and 3 days; in two nests with five eggs each, the time was 3 and 3 days. In a nest of three eggs in which one did not hatch, two hatched on the same day; in another nest with four eggs, in which one did not hatch, the remaining three hatched in 2 days. Six of the females hatched one egg on the first day; seven hatched two eggs the first day, and one hatched three eggs the first day. It was not necessarily the largest clutches which produced the most nestlings on the first day of hatching. Only one egg hatched on the first day in nest 5L; two more hatched on the following day, and the last two on the next day. The small samples available show no important differences in time between the early season broods, and those in later months.

II. The Role of the Male: Secondary Nests

The literature on the Cactus Wren is filled with accounts of additional nests built in the vicinity of the breeding nest. Various terms have been applied to them, such as auxiliary, decoy, dummy, extra, shelter, spare, and supplementary, depending upon the use which the observer attributed to them. In view of the many conjectures and misstatements in the literature, we are reporting in considerable detail the events of the year 1958, and we are summarizing our data for 13 of the annual nest building sequences that we observed from the laying of the first egg to the fledging of the last young or the nest's abandonment. The data for the remainder of the years listed in Table 9.1 are too incomplete to be included here.

Normally, while the female incubated her eggs in the new breeding nest, the male began construction of one or more secondary nests. In Fig. 3.5 are shown the location of the numbered chollas in lots 6 and 7 on Kleindale Road in which the following data were secured. Unless otherwise stated, all the nests were in cholla cacti. In order to simplify the accounts, we have omitted our data on nest-building activities of immature Cactus Wrens. These will be described in a later chapter.

Nest building in 1958 was almost entirely in lot 7, and it followed closely the main predictable pattern, with several interesting variations. Three wrens were present at the beginning of the year, HM-73, HF-71, and a noband. The male roosted in nest 5K, the female in nest 1I, and the noband in nest 21G. By 2 January, HF-71 had moved into the male's roosting nest, 5K, in the northwest part of lot 7 and laid her first egg. HM-73, forced to find another nest, did not take over the female's roosting nest, but began work on nest 21H on 1 January and gradually completed it. The noband wren which occupied nest 21G, only 6 feet (1.8 m) away, was not seen after 10 January. The secondary nest, 25G, located 20 feet (6 m) from the breeding nest, was begun on 14 January. It appeared finished on the 24th. On 9 February, HF-71 carried some lining material to nest 25G; HM-73 also worked on this nest. Fledging occurred on 11 February. HF-71 squeezed into her old, weathered nest 1I that evening, for nest 5K was again crowded with the returned fledglings. She laid the first egg of her second clutch in nest 25G on 15 February. The next day, HM-73 began work on nest 83C in the south part of lot 7. He had given up his roosting nest to the fledglings and now retired temporarily in nest 5K. Later he occupied nest 83C. On 1 March this nest had a slightly torn entrance. HM-73 started 21I but soon abandoned it.

The secondary nest, 5L, 14 feet (4.2 m) from nest 25G, was not started until 15 March; it was well outlined on 21 March, but lacked the interior lining. Nest 91G, found 11 March partly completed, probably was begun earlier and was left in favor of nest 5L. HF-71 apparently roosted in nest 5L on the nights of 23 and 24 March. Fledging from nest 25G took place on 25 March. Two days later the first egg of the third clutch was laid in nest 5L. HM-73 completed nest 91G on 30 March and roosted in it; by 10 April he had repaired nest 83C also. This nest fell apart soon and was again repaired. Now the male gave up his roosting nest, 91G, to an immature wren and retired in nest 83C. Finally, on 4 May, HM-73 began construction of secondary nest 25H, just above the old breeding nest 25G. Fledging of the nestlings in 5L began the following day. The male worked rapidly on the new nest. The first egg of the fourth clutch was laid in nest 25H on 8 May. On 11 May, HM-73 began work on nest 5M, near nest 5L, and again worked rapidly. On 22 May we discovered that he was roosting in a new nest, 96D, on Flanwill Street. Three of the eggs in nest 25H hatched, but on 29 May the nestlings had disappeared. After this failure, the adults began construction of nest 17J in the south part of lot 7 on 31 May. HF-71 roosted in nest 5K, HM-73 in nest 96D. On 5 June the first egg was laid in nest 17J. HM-73 began work on 7 June on secondary nest P4 in the large pyracantha bush beside our front door. Then he shifted his labors to another nest, 92D, near the northwest corner of our house, but he soon abandoned this one also. Incubation continued in nest 17J, but the eggs proved to be infertile. The nest was not occupied on 29 June. The next day, HM-73 and HF-71 joined in building nest 92D, which the male had tentatively started on the 10th. On 4 July the first egg of the sixth clutch was laid. The secondary nest, P5, located in the top of the pyracantha bush, was begun by HM-73 on 8 July. He worked slowly on it, while he roosted in nest 96D. On 3 August the infertile eggs in nest 92D were abandoned. The course of nest building in this most productive year is diagrammed in Fig. 11.1.

Secondary nests were occasionally begun as early as the day following the laying of the first egg of the first clutch. Normally we could expect to find them under construction from 8 to 14 days later. Destruction of roosting nests contributed to delays in starting these secondary nests. Greater delays sometimes occurred when the male was occupied in feeding a hungry group of three to five fledglings. Yet, even then, he some-times managed to start a nest shortly after a clutch was complete. Nest 25H, in 1958, was begun the day before the nestlings were fledged from nest 5L, but the male had previously worked on two other nests. Milam Cater (personal communication) saw a male Cactus Wren begin a secondary nest on 26 May 1943; the following day this nest appeared completed. On the 28th another nest was begun by the male and finished by 30 May, when the young were fledged from the breeding nest.

It was not unusual to discover that a secondary nest had been built in the same cholla which contained the breeding nest. Some nests were only a few feet away; others were back to back or a little above the primary nest (Fig. 11.2). Locations elsewhere in the territory naturally depended upon the availability of nesting sites in cholla cacti. Distances from the breeding nest varied from 14 to 240 feet (4.2 to 72 m).

Secondary nests were well covered over and were sturdy and sub-stantial in appearance in from 4 to 14 days. They often required more lining and additions to the entrance when the female moved in; the work of the male had usually been interrupted by the necessity of assisting the female in feeding the nestlings. Only once did we see the female take any part in the construction of a secondary nest which the male had begun. After the fledging of her young, both adults finished the new breeding nest.

We have records of three temporary secondary nests which were built by females at the time their breeding nests were too crowded with

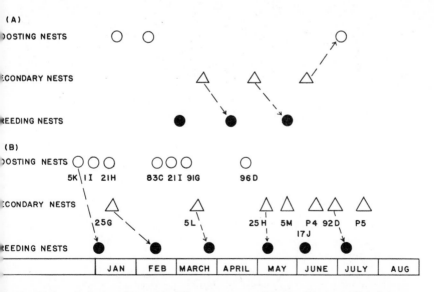

Fig. 11.1.   A, "normal" nest-building sequence of Cactus Wren on Kleindale Road, Tucson, Arizona. B, 1958 nest-building sequence showing variations.

Fig. 11.2.   Cholla 17 on 1 March, 1959, Kleindale Road, Tucson, Arizona.
Roosting Nest 17J is at upper left, breeding Nest 17L at upper right,
and secondary Nest 17M at top. Note the characteristic drooping joints
of the cholla caused by lack of rain.

nestlings to permit further night brooding. These nests were small and
flimsy, but they served as roosting nests until the young were fledged.
Two of the nests were built just above the breeding nest; the third was
at a distance of 160 feet (48 m) from the breeding nest. In each of these
three instances, other nests were available for roosting, but for some
reason they were not chosen.

Under undisturbed conditions, a secondary nest was always built
by the male while the female incubated her first clutch. With but two
exceptions, this nest was a new nest. Nest 23F, in 1945, and nest 17M,

in 1959, were started earlier in the year by the females and then abandoned. Later they were completed by the males.

In the course of the 13 breeding seasons, the male Cactus Wrens began construction of a total of 59 secondary nests while the females were incubating, brooding, or feeding their nestlings. Some of these nests were not completed. If we omit the 23 nests which were begun during the last clutch of the year, and which obviously could only be used for roosting nests, we have 36 secondary nests available for use as breeding nests. Fifteen of these were chosen by females to be their breeding nests. The histories of the remaining 21 nests vary. Several were begun and then abandoned for no reason apparent to us; others were damaged by thrashers, and then a new one was started elsewhere. Some were certainly intended for roosting nests, since they were begun immediately after the destruction of the male's roosting nest. Such a nest was usually built hurriedly, with a secondary nest following before the young were fledged. Whenever the male, apparently willingly, gave up his roosting nest to the still dependent fledglings, he constructed another roosting nest for himself. We feel safe in stating that all completed secondary nests were occupied at some time by the male, the female, or the young wrens. There were no nests which could be considered as decoys, and none which was superfluous. The rejection of a secondary nest by the female did not necessarily mean that there had been a waste of building time, for the fledglings soon needed enlarged roosting quarters. In fact, it is hard to avoid attributing to the Cactus Wren the ability to plan ahead!

Much more research is needed to determine what physiological requirements, if any, dictate the construction of roosting nests. It may be that such nests are not at all necessary in this mild climate. The Curve-billed Thrasher in the same spacial environment as the Cactus Wren, and a direct competitor for nesting sites, is able to hold its own, side by side with the Cactus Wren, without the aid of roosting nests. If roosting nests are necessary for the Cactus Wren, then it seems certain that secondary nests must have an important survival value.

Our relatively infrequent visits to the Santa Rita Experimental Range, and the time required for a thorough population study in the Saguaro National Monument, left little opportunity for investigations of secondary nests in those areas. That they were constructed was obvious, but we did not attempt to catalog them or distinguish them from the breeding nests.

12. The Nestlings

The first year of a Cactus Wren's life can be divided into four periods. We define them as follows: A *nestling* is a wren still in the nest. When it leaves the nest for the first time, it becomes a *fledgling,* and it remains in this category until independence is reached. From then on, we call it a *juvenile,* until the postjuvenal molt in the fall of the year is over. After the completion of this molt the wren is an *immature,* but it is indistinguishable in the field from its parents.

From the south window of our home on Kleindale Road we had a convenient view of the Cactus Wrens feeding their nestlings in nest 19C. We had found the first egg to be hatched at 0735 on 13 April 1947; by 1720 another had hatched. When our continuous observations began at 1750, HF-50 entered the nest with an inch-long (2.5 cm) insect with folded wings. She stopped in the entrance, and her tail could be seen bobbing up and down as though she were hammering at the insect in order to break it up. She then went inside and brooded for 13 minutes. At 1807 the male entered, and departed at once. Five minutes later, HF-50 carried another insect inside the nest.

At 0727 on the following day we found 3 nestlings. The fourth egg did not hatch. The male carried a worm about 0.25 inch (16 mm) long into the nest just before we inspected it. Our first observation period began at 0821 and ended at 0847. HF-50 brought food to the nestlings three times, and brooded twice for 8-minute intervals. She was away from the nest 4 minutes each time. While she was brooding, the male visited her three times. The food he brought may have been given to her instead of the nestlings.

Resuming our watch at 0958, we saw HF-50 enter nest 19C with food. She left almost at once. Then at 1012 HM-54 arrived with what looked like a brown spider. He went inside, but came out in a few moments, still holding the object. Then he swallowed it and flew. Evidently there was no begging response at his entrance, so he could not dispose of the food. At this point there would seem to be a natural transition developing for the male, from courtship feeding to nestling feeding. The female took time out for 22 minutes. Next, she brooded for 9 minutes. Four minutes later she brought food again; then she brooded for 10 minutes. In the course of the next 13 minutes while she was away, the male brought food to the nest three times.

Our third observation period was from 1530 to 1610. At 1535 HM-54 carried a small black insect to the nest; the female was absent. Again he must have failed to elicit a begging reaction, for he came out holding the insect. This time he flew away with it. HF-50 then came with food; she brooded 4 minutes. Eight minutes later, HM-54 returned with a small worm; he visited the nest again in 7 minutes. The female was absent about 25 minutes; then she brooded for 5 minutes.

We watched the nest again from 1830 to 1900, when HF-50 retired for the night. The male brought food twice, the female four times. Her brooding periods were now very short: 3, 1, 1.5 minutes; her periods away from the nest were 9, 3, 12 minutes.

On 15 April 1947, from 0816 to 0908, HF-50 fed the nestlings five times; the male came with food seven times. The female's brooding time was 6, 6, 11 minutes; her time off the nest was 5, 18, 6 minutes. On several of his visits, the male found the female inside the nest. The disposition of his food is doubtful again. He may have fed his mate. From 0926 to 1030 HF-50 brought food five times; the male came three times. The female's brooding periods were 0.5, 0.5, 10, and 2 minutes; her time off the nest was 4, 3, 22.5, 16.5, and 5 minutes.

These incomplete data for 14 and 15 April indicate that approximately 30 percent of the time was devoted to brooding the young. The brooding intervals became very short in the late afternoon of 14 April, and they were even shorter in midmorning of the 15th. Morning minimum temperatures were 42F (5.6C) on the 14th and 47F (8.3C) on the 15th. The afternoon maximum rose to 90F (32.2C) on the 15th. Even though body temperature regulation had not yet begun, the nestlings must have had little difficulty in keeping warm with a minimum of brooding by the female.

On 16 April brooding practically stopped after 0900. The time that the female remained in the nest was hardly more than would be necessary to insure that the nestlings were properly fed. Occasionally a 3-minute period would be recorded, but most of the feeding periods could be expressed in seconds.

On 19 April 1947, the last-hatched nestling was 5 days old. We were able to obtain the rate of feeding for the entire day, from the first visit to nest 19C at 0538 to the female's retirement at 1912. The hourly rates for both sexes in the course of this day are recorded in Table 12.1.

Table 12.1. Hourly feeding rate at nest 19C on 19 April 1947.

Time	HF-50	HM-54
5:38 to 6:38	9	4
6:38 to 7:38	3	4
7:38 to 8:38	7	8
8:38 to 9:38	5	2
9:38 to 10:38	7	5
10:38 to 11:38	7	3
11:38 to 12:38	2	1
12:38 to 13:38	9	6
13:38 to 14:38	7	4
14.38 to 15:38	8	4
15:38 to 16:38	10	3
16:38 to 17:38	7	4
17:38 to 18:38	13	8
18:38 to 19:12	5	2
Total visits with food	99	58

The sky was clear all day; we had light intermittent winds during the afternoon. The minimum temperature was 50F (9.9C), the maximum 90F (32.2C), and mean 70F (21.1C). HF-50 averaged 7.3 visits per hour; HM-54, 4.3 visits per hour. Brooding periods were observed three times in the early morning, of 16, 11, and 5 minutes' duration. Inattentive periods, while seemingly very irregular in length, varied roughly to peaks 12 times during the day. There were 23 inattentive periods of 10 minutes or over, ranging from 10 to 30 minutes in which no feeding occurred. After each long absence or group of long absences, there followed a series of short ones, as the feeding rate increased.

Wheelock (1904: 278) reported that nestling Cactus Wrens were fed by regurgitation for the first 4 days. We have found no confirmation of this anywhere in the literature; in fact, there appear to be no direct observations at all on the early feeding procedure in any recent publication. This is not surprising, for nothing can be seen of the actual feeding after the adult enters the nest. We described the feeding activities in every detail in an endeavor to show that the initial food of the nestlings is not delivered by means of regurgitation. The food consists of small, freshly-killed insects. Our male, when he brought food to the nest, probably gave it to his mate, if he found her inside. At other times, when the nestlings met him with open mouths, the natural thing to do would be to place the insect into the trembling, bright-colored cavity. Both adults brought visible, whole insects to the nest. The female could have chewed up her food while she brooded and then fed the nestlings, but sometimes the brooding period was far too short for this. The difficulty experienced by the male when he brought insects to the nestlings which were too large for them to swallow indicates that no mastication with subsequent regurgitation was contemplated. It was a case of take the food in its original condition or do without it. Our observations at a number of nests confirm the fact that small insects are carried in the bill of the adult to the newly hatched young on the first 3 days of their lives.

Flights to the nest were directly to the doorstep with no attempt at concealment. Flights from the nest also were apparently directly to the feeding areas, mostly east, south, and west. Occasionally these flights were north over our house, but they did not go far in this direction, since it meant trespassing on another territory.

We observed the recognition display-growl at the nest four times in the forenoon on 15 April. Three times it occurred when the male arrived at the doorstep and met the female coming out of the nest. On the fourth time, the male uttered the growl as the female alighted in front of him when he came out. This may have been a defensive or aggressive reaction as he was taken by surprise. Apparently HF-50 did not respond to the display, for HM-54 failed to move aside; she squeezed by him with some difficulty and crept into the nest with her food. By 19 April, feeding of the nestlings had become a routine activity. If the male arrived when the entrance was blocked, he waited and stood aside just enough to permit his mate to leave. There seemed to be no sign of recognition or interest. At one of these visits he gave his food to HF-50, who turned around and carried it inside the nest. It all appeared so mechanical. In the course of the entire day, not one display or growl occurred when the adults met at the entrance.

Singing by the male was more frequent during the forenoon of 19 April. It tapered off in the afternoon and diminished rapidly after 1800. It occurred usually after one or the other of the Cactus Wrens had fed the nestlings and left. Some of the songs may have been in response to activity in adjacent territory I. Few songs, however, were noted during the long periods in which the adults took time to feed themselves. HF-50 sang a number of times at the nest entrance or near it when an immature Curve-billed Thrasher began digging in the ground beneath the cholla. Sometimes her song sounded very much like that of her mate; now and then the syllables were farther apart and possibly were pitched higher. The scratchy *scri* note was heard occasionally during the day, but we did not realize its significance until later, when we discovered that territorial disputes were taking place just north of our house.

The installation of lining material, which occurred so frequently during incubation, was seldom observed after the eggs hatched. At 0756 on 19 April, HF-50 carried some fine grasses to the nest; at 0915 she brought feathers. In neither instance did she remain to brood.

In the course of our limited observations up to 19 April, we did not observe that any fecal sacs were removed from the nest. We presume that they were swallowed by the adults. On the 19th, we saw the adults carry away sacs at the following times: 0558, 0613, 0736, 1046, 1300, 1305, 1306, 1629, 1640. These observations fall into four groups spaced roughly 3 hours apart. Both sexes took part in the disposal of these sacs. On the 25th, HF-50 dropped a large fecal sac as she came out of the nest. The sac broke into fragments; then the female picked them up one by one and swallowed them. Our observations at other nests show that fecal sacs are removed from the nest up to the hour of fledging, and they are often carried as far away as 150 feet (45 m).

Ricklefs and Hainsworth (1969: 34) reported that during the hotter parts of the nesting season the fecal sacs were "often retained" in the nest. In the course of a 12-hour day the total fecal water content amounted to 4-6 ml. From this they endeavor to show that the temperature of the nest would be lowered, but their experiment revealed only a 1.3C drop, and that occurred when 3 ml of water were placed in the nest at one time. If this 4-6 ml of water were distributed over the entire 12 hour-period, as it normally would be, the effect on the nest temperature becomes insignificant.

When the nestlings in nest 6AJ in territory I were fledged on 23 April, their parents led them into the north half of lot 7. Part of our attention now had to be directed to this group of wrens, leaving only brief intervals for observing the feeding at nest 19C in territory III. The boundary disputes which soon began must have disrupted to some extent the orderly routine at nest 19C.

As the nestlings grew larger they became more active. During their final week in the nest, they climbed out of the nest cavity and into the vestibule where they awaited the arrival of food. While they were being fed, their individual begging notes were uttered so frequently that they merged into a loud, coarse *buzz*. On 3 May, the day before fledging, HM-54 sang with increased vigor in the vicinity of the nest. The nestlings crowded forward until one was outside the entrance when HF-50 came with food at 0840. A warning *tek* note from the male sent the nestling back into the nest, but it did not stay there long. Soon two nestlings were out. The adult that came with food poked its head under one of the obstructing nestlings, raised its head and then pushed its way into the nest, evidently to feed the remaining one. It was difficult to keep them inside. For a while one of them squatted in the entrance, eyes closed, apparently asleep. Again they were wide awake, stretching their necks and peering about from the doorstep. At 0941 one of the nestlings climbed to the top of the cholla; it came down part of the way when HF-50 arrived with food. The other two in the entrance were fed. HF-50 moved upward and sang just above the nestling, apparently in an attempt to induce it to come down. This nestling was not fed until 20 minutes later, when it had returned to the nest. No further excursions occurred up to 1030, when we were forced to discontinue watching. For a short period the delivery of food had been at the very rapid rate of one visit per minute. Fledging probably took place some time after 0700 on 4 May. (We were absent most of the day.) On the 5th the three fledglings spent the entire day in the large mesquite tree 20 feet (6 m) southeast of cholla 19.

We watched the fledging at nest 25G on 25 March 1958. At 0900 singing had increased noticeably. HM-73 perched on a dead branch of the cholla near the nest. A nestling came to the entrance and looked around; then it retreated inside. This was repeated several times. HF-71 arrived with food, but instead of going to the nest, she stopped 2 feet (.6 m) away. Soon she flew to a nearby creosote bush; then she came back. Again she flew and returned; then she flew to cholla 4, about 15 feet (4.5 m) to the south. The nestling, which had been watching from the doorstep, fluttered about a foot forward and landed on a cholla twig. It stayed there a few moments, balancing awkwardly among the sharp spines; then it flew downward to another twig. It hesitated and then flew upward another foot and teetered and hesitated again. The rapid singing continued from cholla 4. Finally the young wren flew to the base of this cholla in a curved, descending flight. A few minutes later a second nestling appeared in the entrance of nest 25G. It, too, advanced and retreated on the doorstep several times. Singing continued vigorously. Then the second nestling, without any intermediate stops, suddenly followed the first in a direct flight to the base of cholla 4. A third nestling then came to the doorstep. but it did not attempt to fly. The fourth also remained inside, and both were fed at the nest during the day. By 1030 on the following day, all nestlings were fledged.

It seems evident that at this period the song of the male takes on a new function. The singing, which occurred immediately after the feeding of the nestlings, is difficult to explain. It was not usually uttered in response to the territorial song of adjacent rivals; it seemed to come spontaneously. At the time of fledging, however, the song appeared to be directed at the nestlings, and it served apparently as a signal for them to leave the nest and fly toward the singer. In the evening, the same frequent singing led the fledglings back to their nest to roost for the night. The only difference that we can detect between this song and territorial song is the more rapid rate of the former. There are more songs per minute; the shorter pauses permit little time for listening to any other songs.

The length of time which the nestlings spent in the nest, calculated from the day the first egg hatched until all the young had been fledged, varied from 19 to 23 days. The average of 13 nests was 20.9 days. The single nestling in nest 14C had been in the nest for 23 days when it left on 21 March 1945. In 1958 the three nestlings of the first brood remained in the nest for 23 days; two nestlings of the second brood left in 22 days, the remaining two on the next day; the five nestlings of the third brood left after 21 days. This does not indicate that late broods spend less time in the nest than do the earlier broods, for we have two records in March of 19 and 21 days. Fledging was accomplished in the course of one day in 14 nests; two nests required two days. Evidently the spread in time of fledging is not determined by the spread in hatching. Hatching was spread over 3 days in nest 5L in 1958, but all of the nestlings were ready to leave on 5 May. Repeated disturbances, such as removing the nestlings for weighing and examination, did not cause early departure, provided they were discontinued several days before fledging. We feel sure that premature fledging occurred occasionally when people approached too close to the nests.

The intervals between the fledging of the young and the laying of the first egg of the second clutch show considerable variation. In 1945, fledging and laying occurred on the same day, but only one nestling was involved. The maximum interval was 13 days in 1940. The average for 7 years is 6.8 days. Data for only 4 years are available for the interval between the second and third clutches. The minimum time was 1 day, the maximum 11 days, and average 6 days. The year 1958, with the pair HM-73 and HF-71, can be summarized as follows: 4 days between the first and second clutch; 1 day between the second and third; 3 days between the third and fourth. After the failure of the fourth clutch, the pair of wrens built a new nest and laid after 7 days. This, too, failed. They built another nest and again they laid the first egg in 7 days. The year 1959 with the same male, HM-73, but with a new female, HF-86, progressed as follows: the interval between the first fledging and the second clutch was 10 days; the interval between the second fledging and the third clutch was 13 days. First laying occurred on 19 February in 1959 as against 2 January in 1958.

The time required for a successful nesting, from the laying of the first egg to the fledging of the young, averaged 38.4 days for 14 broods. The minimum time was 36 days, the maximum 41.

The environmental, physiological, or psychological factors responsible for the termination of the breeding season are difficult to determine. Our data for 14 years, obtained from 16 territorial pairs on Kleindale Road, are striking in their variability and afford few hints for any conclusive answer. Three pairs of Cactus Wrens ceased breeding in May, four in June, two in July, six in August, and one in September. The season, measured from the laying of the first egg to the fledging of the last young, or the abandonment of the nest, varied in length from 3 to 7 months, with the average at 4.4 months. This average appears to be an unsatisfactory figure, for the minimum length of season occurred when one or the other of a pair was lost. The acquisition of a new mate in the middle of the season did not always result in the initiation of another breeding attempt. Human activities in the neighborhood no doubt disrupted nesting attempts at times. Perhaps the gradual restriction of the wrens to nesting sites in lot 7, where they were protected, is a factor here. Allan R. Phillips (personal communication) informed us that on 20 September 1938,

Cactus Wrens were fledged from a nest near his home in the thickly populated university district of Tucson. Hensley (1959: 90) found that in Organ Pipe Cactus National Monument in southwestern Arizona, the 3 months' breeding season extended from late March to late June. Since the late June date (Hensley, MS) refers to eggs laid and not to young fledged, the season in that area probably ended in August. We suspect that in the Tucson region the normal physiological decrease in sexual activity begins in the first half of July, after the daytime maximum temperatures have been maintained for several weeks at from 100 to 110F (37.7 to 43.3C). Eggs laid in early July produce young that are fledged in August. In 1955, 1956, 1957, and 1958, laying began respectively on 12 and 7 March, 21 February, and 2 January. The breeding season was over on 7 September, and 10, 13, and 3 August. The longest breeding periods occurred when the human population was increasing in the neighborhood.

Dates of fledging could not be determined accurately on the Santa Rita Experimental Range. We can assume that the late start of laying must have shortened the season considerably. In the presumably normal, undisturbed conditions of the Saguaro National Monument the nesting season, counting successful nests only, from the date of the first egg laid to the fledging of the last nestling — not necessarily from the same pair — extended to 116 days in 1963, 91 days in 1964, 118 days in 1965, 126 days in 1966, 87 days in 1967, and 179 days in 1968. The average length was 3.9 months. The length of the season for six double clutches in 1963 averaged 90 days; for four double clutches in 1964, 85 days; for seven double clutches in 1965, 95 days; for two double clutches in 1966, 85 days; and for three double clutches in 1967, approximately 80 days. A successful triple clutch in 1966 required 126 days, another in 1968, 135 days.

Conjectures on what factors terminated the season of laying in Saguaro National Monument are no more fruitful than they were for the earlier Kleindale Road studies. The average date of the last fledging in the years 1963 to 1968, inclusive, is 12 July; it varied from 6 to 20 July. These dates coincide roughly with the advent of the summer rains in the middle of July, and arouse the suspicion that somehow rainfall caused a cessation of breeding. This suspicion, however, would appear to be discredited by the fact that a warm "wave" preceded by adequate rainfall apparently triggered the start of laying in the spring. The exceptionally long season of 1968 resulted not only from the early laying (24 February), but also from the late fledging of 21 August. There is only a 5-day difference in the average date of the second clutches of 1965 and 1966. In 1965, laying ended with the second clutch; in 1966, it continued for a third. The influence of any temperature departure from normal in the consistently hot and dry month of June appears insignificant and negligible. Cloudy, threatening skies form in the last half of June, and frequent thunderstorms occur in July. Third broods were often begun in the first week in June, some time before there is a hint of the rain which precedes the abundant growth of summer ephemeral plants and their associated insect life. Nice (1937: 135) attributed cessation of nesting to the drought which brought on premature molt in the Song Sparrows. In the Monument, nesting proceeded often under drought conditions in April, May, and June, while the annual plants completed their cycles and dried. The larger perennial plants, drawing upon their winter rainfall storage, flowered later and made up some of the deficit in food supply in the driest months. It is possible that the age of the female is a factor, but unfortunately in 1965 only three of the 19 females had bands. Two of these females were missing in 1966, and the third laid only two clutches. The remaining eight females consisted of five nobands, one banded year-old (she laid three clutches), and two banded ones of unknown age.

13. Development of Nestlings

Our data on nestling growth and development of the Cactus Wren, from time of hatching to time of fledging, were obtained from five nests, two in 1940, one in 1941, and two in 1958. The nest studied in 1941 was a first brood of the year; the others were first and second broods. We visited the nests near our home, in the Kleindale Road study area, once a day, usually between 1700 and 1800. We removed all the nestlings of a nest at the same time and carried them to our house, where we examined them carefully and weighed them to 0.1 gram. Immediately thereafter, we replaced them in their nest. If the adult birds observed us, they made little protest unless the young squealed. Sometimes they must have been entirely unaware of our intrusion, for on some such occasions they were not seen at all. Our interruptions of their feeding routine seemed unimportant so far as weight of the nestling is concerned. In the following accounts, the nestlings are considered to be a day old on the evening of the day after hatching. Although this introduces a possible error of an additional 12 hours, the actual error is probably much less. Some eggs hatch in the course of the afternoon.

The descriptions which follow roughly represent averages of individual variations in development. Although the sequence of feather development appears to be the same in these nestlings, the time at which changes occur may vary as much as 48 hours. As is customary in these measurements, we have used metric units throughout.

Age 0 days. At hatching, the eggshell is cut in a ragged line almost exactly around the equator. The separated halves may adhere for a brief time to the extended head and the swollen abdomen. When freed from the shell, the nestling is still wet; the down sticks close to the skin in irregular streaks. Later, when it dries, the down becomes whitish and fluffy. It varies in length from 4 to 8.5 mm and is present in tufts in the capital, spinal, humeral, alar, femoral, crural, and ventral tracts (Fig. 13.1). There is considerable individual variation in quantity of down in members of the same brood. The bare crown is usually surrounded by a ring of down which extends backwards from the forehead, passing above the eyes to cross the occiput. Sometimes this circle is incomplete. There is a dense tuft in the dorsal region, and usually a smaller one in the pelvic region, just above the oil gland protuberance. From one to five small tufts have been found on the humeral tract, and as many as seven in the alar region; as many as eight have been seen in a row in the femoral region. Fewer tufts are present on the front and back of the crural tract. There are from one to three small tufts on each side of the lower abdomen; otherwise the ventral tract is bare. Sometimes the skin of the cervical, dorsal, and crural areas appears wrinkled.

Saunders (1956: 122) has assumed that the presence of natal down on only the capital and dorsal tracts is a character of the wren family, but his observations were limited to only two species, the House Wren and the Long-billed Marsh Wren (*Telmatodytes palustris*). This assumption proves invalid when the Cactus Wren is considered.

The bill is pink, above and below; the swollen edges are yellow from the angle of the mouth to a point about halfway along the length of the bill. Near the end of the slightly shorter upper mandible the egg-tooth projects upward as a tiny, light-colored, sharp point, less than 0.5 mm long. The closed eyes appear as large slate-gray bulbs with a suggestion of a slit in the lower, paler portion of the surface. The mouth lining is orange-red.

Age 1 day. Very little change, except for a slight darkening of the skin.

Age 2 days. The mouth lining changes to orange-yellow; the crown becomes darker, a dull reddish color, with small, indistinct specks in the skin. A dark streak has appeared along the center of the manus and small dark specks can be seen on the forearm. In the caudal tract, tiny bristles indicate where the tail feathers will appear. The skin of the spinal and ventral tracts is darker, and all tufts of down are gray.

Age 3 days. The bill is pinkish-brown now; the eye-slit, still closed, has become more prominent, and the crown, now dark reddish black, appears wrinkled and speckled. In the alar region, which is also darker, the primary and secondary bristles are visible. A dark area can be seen just anterior to the feather germs of the caudal tract. The skin of the cervical tract is darker than previously. Faint spots can be seen on each side of the ventral region.

Age 4 days. The bill has become light brown, but the swollen edges are still yellow. As the head grows larger, the eyeballs protrude less and

Fig. 13.1.   Nestling of Cactus Wren, Kleindale Road, Tucson, Arizona
From top to bottom, ages 0, 7, 11, and 18 days

the eyelids appear almost black. On some nestlings the eye-slit is cracking open very slightly. Two-thirds of the outer surface of the forearm and manus is blackish, and the primary and secondary bristles are 0.5 mm long. A dark area has appeared under the skin of the humeral tract. The dark caudal area is more prominent, and all of the spinal tract is darker and dotted. In the ventral region the light sheaths beneath the skin outline the entire tract; the axillary branch can be seen for the first time.

Age 5 days. All the original down is still present. The bill is light brown at the tip and paler around the nostrils. The glint of the eyeball can be seen through a slit now open for about 1 mm. Light-colored sheaths are apparent under the skin of the femoral and crural tracts. Most of the outer wing surface is blackish; the sheaths of the primaries, secondaries, and their coverts have pierced the skin. The humeral sheaths are still beneath the skin. A dark semicircle has appeared anterior to the light-colored papilla of the oil-gland. The points of the sheaths in the entire spinal and ventral areas are visible beneath the skin.

Age 6 days. The egg-tooth has become almost microscopic; in some nestlings it is entirely gone, but a small bump will remain for at least a week. The eye-slit is open 1.5 mm; the width of the opening is so slight that it is improbable that the bird can see effectively. Growth has continued in the primaries, secondaries, their coverts, and the alula. The sheaths are breaking through the skin in the spinal area. In some nestlings sheaths have emerged in the ventral and crural tracts.

Age 7 days. The tip of the bill has become a darker brown, and the mandibles appear equal in length. Some vision may be possible, for the eye-slits have opened to 2 mm, and the area between them is oval in shape. The lids are movable, opening and closing the small oval. On the capital tract the sheaths have erupted through the skin on the forehead, crown, and nape. They have likewise emerged in all the remaining tracts (Fig. 13.1). The dorsal tract has fanned out. There are three rows of closely spaced sheaths on each side of the abdomen; at the axillary branch of the ventral tract, seven rows are visible. Widely scattered sheaths are present on the breast between the ventral tracts, and a denser growth is apparent on the chin.

Age 8 days. The tips of the sheaths are breaking open, showing small, light-colored, fuzzy tufts at the ends of the primaries, secondaries, and some of their coverts, and on the feathers of the dorsal, femoral, crural, and caudal tracts. Down is still present.

Age 9 days. The eye-slit opening appears as an oval about 3 mm in length. In the ventral region the axillary sheaths are the first to burst at the ends. The upper mandible has lengthened.

Age 10 days. The eye opening is now almost round and the lids open and close more frequently; rows of sheaths are evident on the eyelids. A whitening of the superciliary sheaths is apparent. Sheaths cover half of the ear opening. Most of the sheaths of the ventral area, except those on the chin, have burst at the ends. The ends of the second-aries are now fan-shaped and about 3 mm wide.

Age 11 days. The eyes are now open wide. Sheaths of the nape and the superciliary stripe have begun to open; sheaths of the greater primary coverts have burst. All other tracts show further widening of the fan-shaped ends of the feathers as the sheaths continue breaking (Fig. 13.1).

Age 12 days. Most of the bill, with the exception of the edges, is now darker in color. The sheaths of the entire capital tract are bursting. All of the upper part of the body appears rather well covered when the bird is crouching. The ventral tracts have broadened, leaving a narrow open strip down the middle. Sheaths cover almost the entire ear opening; the chin and throat feathers are still in their sheaths. Down is still present.

Age 13 days. The capital tract now appears smooth, with new, brown feathers, and the white of the superciliary stripe is prominent. Bristly sheaths are still evident at the base of the bill. The wings cover almost all of the dorsal apteria; a small apron of feathers conceals the oil gland. Natal down can still be seen above and posterior to the eyes and on the humeral and dorsal tracts.

Age 14 days. The bill is dark on top with lighter nostrils; its edges are yellow. The eyelids appear fringed with slender sheaths. When the nestling is at rest, all the dorsal surface appears feathered, and the tips of the wings reach as far back as the oil gland. Only the central abdomen is bare.

Age 15 days. The primaries and secondaries are out of their sheaths for about half their length. Disintegration of the sheaths of the body feathers is so extensive that the broken pieces fall like chaff when a nestling is handled.

Age 16 to 17 days. Feathers cover the entire body with the exception of the lower abdomen. There is an indistinct center line in the ventral region, but all apteria are rapidly being concealed by the spreading feathers. The iris is pale grayish or yellowish. It will be weeks before it darkens and finally becomes red. No further examinations were attempted because of the risk of premature departure of the wrens from their nest, but a photograph was taken at the age of 18 days (Fig. 13.1).

Behavior of nestlings

Age 0 days. The nestling appears very weak and is unable to right itself, although it moves about when handled. Occasionally a bird of this age defecates. The head trembles when the bill is opened wide. The legs are too weak to serve for bracing or for grasping. When the bird is not disturbed, the head rests upon the large abdomen. The small pink body becomes cold quickly.

Age 1 day. Weakness is still very apparent, but the wren struggles longer in an effort to right itself. It assumes the embryonic position when resting and becomes noticeably cold when left exposed for a short time.

Age 2 days. More strength is evident as the nestling struggles to turn upright. When suddenly touched, it responds by moving. Sharp *peep* notes are occasionally heard, especially during handling.

Age 3 days. Although strength has increased, the nestling still rests in the embryonic position.

Age 4 days. The nestling can lie flat on its chin, abdomen and tarsi, but now and then it drops its chin upon its abdomen. One nestling could stand on its horizontal tarsi, using its bowed wings for braces, while it opened its bill. Gaping has become more frequent, while *peep*ing is variable. Some nestlings are exceedingly vocal, others silent. The nestling becomes cold quickly.

Age 5 days. By this time there is much individual variation in restlessness. Some nestlings lie quietly, grasping one's hand with their claws; others lie flat and try to crawl forward, using both wings and feet. It is difficult for them to remain upright. The head trembles on the outstretched neck.

Age 6 days. Temperature regulation must be at least partly effective now, for the nestlings cool off much less rapidly. Their activity has increased considerably. They are harder to remove from their nest; they grasp the lining with their claws and hold on. The righting reaction takes place with less difficulty. When placed on its belly on a smooth table, a nestling tried frantically to obtain a foothold by "swimming" with its wings and legs. It showed no fright.

Age 7 days. *Peep*ing notes are heard frequently when the wrens are taken from their nest. Their bodies feel very warm. When handled or held, most of their efforts were directed toward assuming a head-up position.

Age 8 days. The nestlings try to crawl away, but no real fright is apparent. This may be merely an attempt to attain a more comfortable situation, away from annoyance.

Age 9 days. Strength has increased to a point where the nestlings crawl and climb so vigorously that they must be placed in a box for weighing. They climb with outstretched necks and grasp with their claws, but they cannot yet perch on one's finger. If the hand holding them is dropped suddenly, they raise their wings and tail instinctively to balance. After we tumbled three of them back into their nest, they righted themselves at once so that they faced the opening, their chins resting upon the edge of the nest cavity. Ricklefs and Hainsworth (1968b: 122) determined temperature regulation became evident at 9 days of age and was complete by the 13th day after hatching.

Age 10 days. Fear reactions are not yet evident. The nestlings are stronger and can perch on one's finger, but they have difficulty in balancing. They have begun to turn their heads slightly in order to peer around at nearby objects.

Age 11 days. At this time the first signs of fright are apparent in some of the nestlings. They exhibit the "back-up" posture, in which the head is lowered with the bill held horizontal; then the middle and posterior end of the body are humped up while the wings and tail are lifted as the nestling tries to back up in the palm of one's hand. (Ricklefs [1966: 48] suggests that this "back-up" behavior may also be a response to the satisfaction of the nestling's hunger, for it usually crawls back into its nest after it has been fed.) Defecation before and after weighing becomes more frequent. Some nestlings squealed loudly when we tried to remove them from their nest; they were packed down like sardines, perhaps in an effort to hide, and were difficult to grasp without pinching. When they were replaced in the nest vestibule, they crawled back into the cavity at once.

Age 12 days. Crouching and withdrawing, as in fear, at sudden movements are increasing now; so also is the effort to climb out of the hand, usually upward, in attempts to escape. They are not yet able to stand on their toes, but rest on their horizontal tarsi. If rolled on their backs, they struggle fiercely to right themselves.

Age 13 days. The head movements are more noticeable and blinking is frequent, as the nestling observes its surroundings.

Age 14 days. Escape attempts are the rule. The nestlings crawl ahead, apparently blindly, regardless of obstacles, evidently unable to judge distance or depth. They will crawl to the edge of a table and fall to the floor, happily without injury. Their legs are still weak, but they can travel rapidly on a carpet where a good foothold can be obtained. If a hand is placed in front of the nestling, it climbs over it, falls, and then continues on its way until it finds a dark corner in which to hide. *Peep*ing notes are still heard during the examination. One nestling protested loudly while it was held in the palm of the hand, perhaps at some unexpected movement near by.

Age 15 days. Rapid peering movements of the head are frequent and attempts to escape, assisted by fluttering of the wings, occur. The birds may try to stand on their toes, but usually they settle down on heels and tarsi. Ricklefs (1966: 51) observed preening in a 15-day-old, hand-raised Cactus Wren.

Age 16 days. Restraint of the nestlings is difficult now, for they are able to run on their toes, and they even attempt to fly a few feet if inadvertently released. All struggled vigorously, grasping at anything within reach as they kicked and churned in trying to turn upright. When finally quieted, they held their bills upward at an angle. One uttered a begging note when it heard a squeaking sound.

Age 17 days. The begging note, *dzup,* was heard from three nestlings on this day. It was uttered several times when a wren was placed on the floor alone. One hears this note frequently in the course of the last week of undisturbed nestling life. Its occurrence while being examined would indicate that this nestling was not alarmed at the time. Ricklefs (1966: 51) reports stretching and head scratching in his hand-raised bird at this time.

Age 18 days. The wren photographed in 1941 (Fig. 13.1) uttered a lively *dzip* note frequently while being set up for its picture. Apparently it was not afraid as long as we did not touch it, or move suddenly when near it. On being returned to its nest, however, it escaped and flew a distance of about 15 feet (4.5 m) to a small bush. Here it remained and permitted us to pick it up again. As soon as it was placed in the nest entrance it crawled inside and vanished in the dark, crowded cavity.

On the 19th day, at nest 27B, we pulled out the five nestlings, one by one. Two escaped at once, the first flying a distance of 30 feet (9 m), the second 75 feet (22.5 m), accompanied by a noisy, frantic, adult wren. This nestling took refuge at the foot of a creosote bush, and, by running to the opposite side of the bush whenever we approached, frustrated all our attempts to catch it. We replaced the remaining three nestlings in their nest and abandoned any further efforts to photograph them. On the following day at 1800 the nest was vacant.

The progressive increase in escape reactions after the age of 10 days was very pronounced in all the wrens we examined. Our regular visits, begun on the day of hatching, did not eliminate these escape reactions. In the course of the examination, each nestling was handled for an average of 5 minutes; the total time out of the nest for the entire brood did not exceed 30 minutes. It is difficult to see how these brief periods could have had much effect in altering the normal behavior of the nestlings. Our observations agree with those of Banks (1959: 101) on the White-crowned Sparrow and are sharply at variance with those of King (1955: 160-161), whose study of the Traill Flycatcher (*Empidonax traillii*) indicated that escape reactions did not appear under the routine of daily visits. It is worth noting that the enclosed, covered cavity of the Cactus Wren's nest provides the nestlings with a place in which to hide. Even as late as the 18th day, the nestlings crawled rapidly into the interior and sought safety by concealing themselves from our view. Had these nestlings been in an open nest, it is highly probable that they would have "exploded" into flight several days earlier when we touched them. Premature fledging is, we believe, in most instances actually an escape reaction caused by fright or some threatening disturbance.

Weights

The lightest weight recorded on the day of hatching, 2.6 grams, was that of H-75. This nestling was hatched sometime between 0930 and 1700. It could hardly have been many hours old, for its down was still wet. The greatest weight, 4.1 grams, was that of H-17, but it may have been as much as 24 hours old. It hatched in the course of the night and probably received considerable food before we weighed it at 1740 the following day.

Table 13.1 details the daily weights of 20 Cactus Wren nestlings from five nests in the Kleindale Road area. Table 13.2 shows the mean

Table 13.1. Daily Weights of Cactus Wren nestlings. Nest 6M-1940.

Age	1	2	3	4
0	3.2	3.7	3.9	3.6
1	5.7	4.8	5.1	5.0
2	7.6	6.4	7.3	7.7
3	9.6	8.7	9.6	9.6
4	13.2	11.0	12.3	11.8
5	15.9	14.0	14.1	12.8
6	18.4	16.8	18.2	15.4
7	21.2	20.5	20.8	19.0
8	24.4	23.6	23.5	23.9
9	27.2	25.0	26.8	24.5
10	28.1	28.1	27.0	27.8
11	30.0	27.8	29.0	28.6
12	30.4	28.7	30.4	29.3
13	31.6	30.9	29.3	29.8
14	32.0	30.2	31.1	30.1
15	33.3	32.1	30.0	29.8
16	31.2	30.7	30.9	30.2
17	30.8	31.1		

Table 13.1. Daily Weights of Cactus Wren nestlings. Nest 32A-1940.

Age	1	2	3	4
0	2.9	3.7	4.0	4.1
1	4.5	5.6	6.3	6.0
2	6.7	7.3	8.1	7.6
3	8.8	9.7	10.4	9.5
4	11.0	12.2	12.3	12.2
5	13.5	14.1	14.3	14.2
6	16.5	15.8	15.8	15.7
7	17.8	19.1	19.1	18.7
8	22.5	22.1	21.8	19.5
9	23.7	23.9	19.8	21.4
10	24.6	25.0	22.8	21.9
11	25.6	28.0	24.1	24.0
12	27.7	30.1	27.0	26.3
13	30.2	29.2	27.7	27.0
14	32.5	29.0	26.6	26.5
15	27.8	29.6	27.4	27.3
16	27.8	30.7	28.4	27.3
17	28.8			

Table 13.1. Daily Weights of Cactus Wren nestlings. Nest 27B-1941.

Age	1	2	3	4	5
0	3.4	3.8	3.9	2.9	3.3
1	4.5	5.2	5.5	4.3	4.7
2	6.5	7.6	7.4	5.6	6.9
3	8.4	9.8	10.3	8.3	8.7
4	10.8	13.5	12.8	11.0	12.0
5	14.2	16.1	16.4	13.8	14.1
6	17.6	19.0	19.0	16.6	17.0
7	19.2	21.1	22.0	19.6	20.2
8	23.1	24.8	24.5	22.3	23.8
9	25.0	26.3	26.9	26.0	25.3
10	27.3	27.9	29.1	27.0	25.7
11	28.6	31.0	29.4	28.0	26.1
12	29.3	30.4	29.5	27.9	27.8
13	29.5	29.6	30.1	28.0	28.0
14	29.4	30.4	29.5	28.4	26.9
15	30.1	30.4	28.4	28.1	26.2
16	29.3	29.6	28.2	27.8	
17	29.6	30.0			
18	30.3				

Table 13.1. Daily Weights of Cactus Wren nestlings. Nest 25G-1958.

Age	1	2	3	4
0	2.9	2.7	3.0	3.1
1	4.5	3.6	3.9	4.6
2	6.1	5.5	5.8	6.1
3	8.2	8.2	7.4	8.5
4	10.2	10.4	9.5	10.4
5	12.6	13.5	12.4	13.8
6	16.1	16.9	15.5	
7	19.8	20.5		20.0
8			21.4	24.8
9	25.8	26.2	24.3	25.5
10	28.8	29.1	26.0	26.8
11	30.5	31.3	27.4	28.7
12	31.2	32.7	28.2	29.3
13	30.6	32.2	28.8	29.6
14	31.3	33.7	29.4	30.5
15	31.6	32.9	29.1	30.6
16	32.2	32.9	29.4	
17	33.7	32.0		

Table 13.1. Daily Weights of Cactus Wren nestlings. Nest 5K-1958.

Age	1	2	3
0	3.3	2.6	3.1
1	4.5	4.1	4.7
2	6.8	5.7	6.0
3	9.1	7.5	7.8
4	11.8	10.7	10.4
5	14.1	12.7	12.7
6	17.3	16.6	16.0
7	20.7	20.5	19.5
8	23.0	23.2	23.1
9	25.1	26.5	25.9
10	26.7	29.3	28.3
11	28.5	30.6	31.3
12	29.5	33.5	31.3
13	30.0	33.2	32.1
14	28.9	33.0	33.7
15	31.3	34.5	
16			33.2
17	31.5	34.2	

Table 13.2. Mean daily weights of Cactus Wren nestlings.

Age	Sample size	Mean	Range
0	20	3.4	2.6- 4.1
1	20	4.9	3.6- 6.3
2	20	6.7	5.5- 8.1
3	20	8.9	7.4-10.4
4	20	11.5	9.5-13.5
5	20	14.0	12.4-16.4
6	19	16.9	15.4-19.0
7	19	20.9	17.8-22.0
8	18	23.1	19.5-24.8
9	20	25.1	19.8-27.2
10	20	26.9	21.9-29.3
11	20	28.4	24.0-31.3
12	20	29.5	26.3-33.5
13	20	29.9	27.0-33.2
14	20	30.2	26.5-33.7
15	19	30.0	26.2-34.5
16	16	30.0	27.3-33.2
17	9	31.3	28.8-34.2

daily weights of the nestlings from these five nests. The variation for a given age is considerable, but the overlap is least during the first week of nestling life. Later, the use of weight for the determination of age becomes very unreliable; a nestling weighing 30 grams could be from 11 to 17 days old.

The daily increment of weight is fairly uniform up to the 11th day (Fig. 13.2). Then the curve of mean weight flattens and soon falls in the characteristic manner reported in a number of other species. Before the birds fledge, the curve has begun to rise again. The cause of this leveling off and decrease in the daily weight gained has been a subject of much speculation. It would seem to be a problem for the physiologist, for although many variables could be involved, we need first to know the food requirements for body and feather growth, and the energy consumed in temperature regulation at the ambient temperatures normally experienced by the nestlings. Without these quantitative data, solution of the problem seems impossible. Changes in availability of food or rate of feeding, the latter stemming from an assumed temporary parental exhaustion, are difficult to prove. The loss of weight is consistent in practically all of our nestlings in the five broods. Considerable feather growth has already occurred by the 11th day, but more occurs in the course of

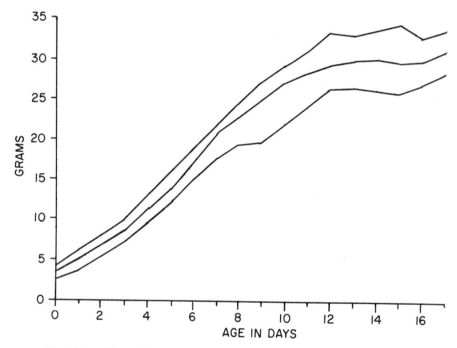

Fig. 13.2.   Mean daily weights of Cactus Wren nestlings on Kleindale Road, Tucson, Arizona. Upper and lower curves indicate range of variation.

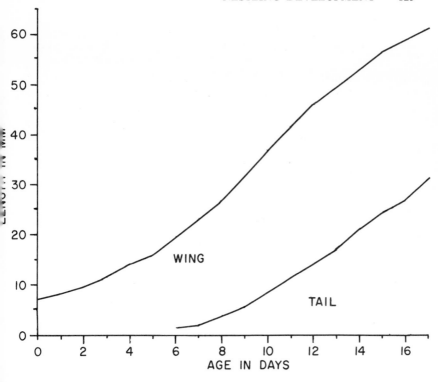

Fig. 13.3.   Mean daily length of wing and tail of Cactus Wren
nestlings on Kleindale Road, Tucson, Arizona.

the next 5 days, so that by the time the bird leaves the nest its short tail
is the only disproportionate element. A small loss of weight takes place
from disintegration of feather sheaths. Weights at time of fledging could
not be obtained, but they were probably about 32 grams.

    Growth of wing and tail. Measurements of wing, tail, culmen, and
tarsus were taken at the time the nestlings were weighed. The last two
measurements proved to be so unreliable that they were discarded. Without feathers for a starting point, the culmen became indefinite; the tarsus
in the live bird could not be held steady long enough for a satisfactory
measurement. The mean daily growth of wing and tail is shown in Fig.
13.3. Up to the fifth day, the rate of growth of the bare manus is rather
slow. Then, when the remiges have broken through, the rate of growth
of the combined manus and primaries rises rapidly to a fairly constant
value and does not begin to drop until after the 13th day. By the 17th
day the rate has decreased to one-half the maximum rate. At the time
of fledging, the wing is probably 75 percent of the adult wing length. The
tail feathers appear to maintain a more steady rate of increase, at least
up to the 17th day, when they have attained 50 percent of the adult tail

length. Both the wings and tail at the time of fledging are of sufficient size and strength to sustain flight, with fair control, for a considerable distance.

Our information on the autumnal, post-juvenal molt is very meager; we did not endeavor to raise any nestlings, nor did we collect any specimens of fall birds. Selander (1964: 12-23) has published an extensive discussion of the molt of the entire genus of Cactus Wrens. He found that in almost all birds some of the juvenal primaries and secondaries are retained, but the tail feathers are replaced.

14. From Fledging to Independence

Weights of fledglings

In the days immediately following fledging, the rate of gain in weight appears to be rather low. It may possibly fall to zero, or there may even be a loss of weight for several days due to delays and postponements of feeding. This occurred often as fledglings dispersed in the Kleindale Road territory and became involved in dangerous man-made situations. In addition, the efforts of the adults to direct their fledglings back to their nest in the evening were time-consuming, and frequently difficult because of human activities.

The low rate of gain in weight which we observed may not be typical. One might suspect that in the absence of disturbances in the original, natural habitat there would be a higher rate. However, the tendency to increase in weight could be offset by the greater difficulty of obtaining insect food in the open, waterless desert.

Wren H-32, who weighed 29.1 grams at the age of 11 or 12 days, was trapped when 25 or 26 days old (5 days after fledging); at that time it weighed only 30.1 grams. H-80, trapped at 37 days, weighed 36.8 grams; H-26, at 38 days, weighed 34.6 grams; and H-78, at 39 days, weighed 40.9 grams. At the age of 54 days, H-76 weighed 41.6 grams. Evidently at the age of about 38 days, 2.5 weeks after fledging, the weight attained is within the range of adult weights. Wing and tail length have also reached adult proportions by this time.

The mean weight of 42 fully grown Cactus Wrens, trapped from 1940 to 1960, including juvenile birds feeding independently of their parents, is 38.9 grams. More than half of these birds were weighed in the afternoon or evening; none of the remainder was weighed earlier than 0850. About two-thirds of the wrens were weighed in the months of August, September, November, and December. The sample is too small to permit any safe analysis of weight variation in the course of the year, or of differences between the sexes. A range in weight of from 33.4 to 46.9 grams was found.

Eleven Cactus Wrens, trapped at their roosting nests after dark, and then kept overnight, lost from 1.9 to 3.3 grams during their confinement, which lasted from 9 to 11.5 hours. The average loss of weight per hour was 0.25 gram. The percentage of body weight lost ranged from 4.6 to 8.1, with a mean of 6.4. This percentage of body weight lost, and subsequently regained, is somewhat higher than the percentage of daily increase of less than 5 percent reported by Nice (1937: 21) for several smaller passerines.

Fledgling and juvenile roosting behavior

Fledging of all nestlings in a nest was usually accomplished in a single day. For the greater part of this day, in a normal, undisturbed habitat, the young wrens were probably never out of sight of the nest in which they had spent the first 3 weeks of life. Those wrens which were fledged on our fenced 1-acre lot seldom traveled more than 50 feet (15 m) from their nest in the first day. The wrens which nested in adjacent lots were subjected to frequent disturbances from human activities, especially in the early evening. Ordinarily, roosting and incubation were not seriously interrupted by the coming and going of people and automobiles. Adult wrens became accustomed to such interruptions. Once the nestlings were fledged, however, the situation changed quickly. Their first day out of the nest was one of extreme danger, if not disaster. The solicitous warning and danger call notes of the parents increased at every human disturbance, and the parents led their fledglings farther and farther away from their nest. Some fledglings were lost because of their failure to return to the nest before darkness.

The adult singing which accompanied the fledging was not reduced noticeably thereafter. It occurred chiefly when the adults were near the young, and usually singing followed feeding of fledglings by adults. The begging note of the fledglings changed from a weak *tek* to the normal *dzip* or *dzup*. Unless frightened, the fledglings were always to be found 2 or 3 feet (0.6 to 0.9 m) above ground in a bush or cholla cactus. Ground travel did not take place until later, when they had learned to follow their parents about the territory as they searched for food. When well fed, the fledglings perched quietly and sometimes appeared to be asleep.

On 5 May 1947, the day they fledged, the three young from nest 19C climbed about in the mesquite tree in our front yard in a follow-the-leader fashion. Some of their movements could perhaps be called play antics; other movements suggested aggressiveness. Once, when two of the fledglings moved upward on an inclined branch, the first one apparently lost its footing and fell. Immediately its companion grasped the wing of the falling bird in its bill and held the bird dangling, while the adults set up a frantic commotion. Soon both fledglings fell into the flower bed beneath the tree and continued squealing from pain or fright. They separated when the parents flew down. A few minutes later, as the same two climbed upward to be fed, a similar encounter took place. This time it may have been a deliberate attack. The fledglings fell scream-ing to the ground, and remained there so long that we felt it necessary to investigate. One of them lay on its back, looking up at the other, which was standing over it. This interruption brought on a demonstration of parental protest; the young squealed as they parted to seek safety in the tree. The male then sang once, after which he flew to his secondary nest to rearrange some straws. His behavior seemed to parallel the dis-placement activities which we recorded in the course of many territorial disputes.

Improvement in flying ability was rapid, for the adults led their fledglings in flight from bush to bush in their territory. After the first day or two of waiting near their nest, fledglings seemed to learn that food could be obtained oftener if they followed their providers. The dangers of landing too abruptly were soon overcome. Branches of mes-quite and creosote bushes offered few challenges. The fledglings landed, teetered, hung on, and then relaxed with decreasing awkwardness each day. Incredibly, the long, vicious spines of the jumping cholla and the shorter spines of the cane cholla were quickly taken in stride as though needles were of no consequence. We never saw a fledgling impaled or even disturbed by these obstructions. Landing, climbing, or walking on a cholla branch seemed to be no more difficult than threading one's way through a short stubble field.

Miller (1936: 218) reported a "mummified juvenal Cactus Wren" impaled on thorns at the entrance of a nest, and Jaeger (1922: 73) men-tioned a similar accident.

In all instances where we had the opportunity to watch the fledglings on the evening of their first day out of the nest, the parents endeavored to lead them back to roost, either to their old nest or to another one close by. Apparently at this time the habit of roosting in a nest must be learned. The fledglings did not return to their nest without parental urging and assistance.

The evening retirement ritual usually began about half an hour before sunset and required from 8 to 10 minutes for completion. At first, the adults flew to the nest in which the fledglings were to roost. Singing increased in frequency as they waited on the doorstep. As if to indicate what was wanted, the parents went in and out of the nest several times, singing rapidly when they came out. Finally one of the fledglings flew waveringly to the cholla, landing in the vicinity of the nest, or sometimes even upon it. Then the other fledglings came, one after another. If they found a footing below the nest, they had little difficulty in climbing upward and squeezing between their parents into the nest. Those on top of the nest, however, appeared unable to muster sufficient courage to move downward. Time after time the parents climbed up to the fledglings and down again into the nest. Eventually, after much coaxing, the descent was accomplished. Sometimes the fledgling actually fell down to the doorstep; not infrequently it tumbled upon the back of the adult bird. By this time, one or more of the other fledglings could be seen crowded into the enlarged entrance, where they faced outward. The last fledgling often had to lower its head and crawl under the other fledglings to reach the inside of the nest. A moment later, its bill and eyes found an opening from which to point outward. Then the male or female followed the last fledgling into the nest, by pushing an opening between two others. The entire brood remained in the entrance, while the parents resumed their regular schedule of feeding. Not until late dusk did the fledglings turn about and huddle in the dark interior of the nest.

The blocking of the nest entrance by the adults is difficult to interpret. We have seen many a fledgling come to a sprawling stop upon the back of one of its parents. After disentanglement, it had to "elbow" its way inside the nest.

The retirement technique of the fledglings gradually improved. They soon learned to fly accurately, so that they alighted on the doorstep of the nest. Later, from more distant parts of the territory, they were led to their roosting nest in short flights from one bush, cholla, or fence post to another. It seems probable that for the first few evenings it was not necessarily the nest, but the song of the adults, which attracted the young. When the adults stationed themselves at the nest entrance, the nest became the destination; and when the fledglings arrived, they recognized the familiar surroundings. It was only a step further to enter a grass-lined opening in which they formerly had found safety in darkness. At this early age, recognition or memory of a nest probably consisted of an opening or a tunnel into which to retreat when feeding ceased and darkness began.

Evidently the fledglings learned rapidly that they were to roost in a nest, for once they were led near one, they crawled in quickly. Nevertheless, the process of daily parental instruction and assistance continued for at least 17 to 25 days. Even after the young became independent, as long as the group remained together, the juvenile wrens returned with their parents to the roosting area. At first the young wrens appeared to have no memory at all of the location of the recent home, or at least no inclination to return to it. They entered any nest in the evening if the adult wrens sang from the doorstep.

How do Cactus Wrens know when all of their brood has been assembled at the roosting nest? Can they count? Probably not. In the course of the first day or two out of the nest, the fledglings were within sight of their roosting nest, and no difficulty should have been experienced by their parents in seeing them. Later, as the area of dispersal increased, a visual, inclusive count of fledglings by adults might become impossible. Perhaps the adults could sense that more than one fledgling should be in the nest; but after two or three fledglings entered the nest, the adults might not know for sure if there were others, unless the fledglings were heard begging. As long as the fledglings could be heard in the vicinity, the work of leading them to roost continued, provided darkness did not intervene.

Retirement before dark is evidently essential. The adults waited for the fledglings only a short time after sunset, for twilight at this latitude is relatively short. If the fledglings were still out of the nest at that time, only a few more attempts were made to call them in. If these attempts were unsuccessful, the adults retired into their own roosting nests, abandoning the fledglings wherever they happened to be. In May 1940, on the evening of their first day out of the nest, three fledglings from nest 32A were found to be 400 feet (120 m) northeast of their nest. Apparently they had been disturbed so severely that their parents were unable to direct them to their roosting nest before dark. The fledglings spent the night in a mistletoe clump in the top of a catclaw bush. On the following day the adults located them and resumed feeding, but in the evening the adults again failed in attempts to lead them back to their nest. We never saw the three fledglings again.

The fledglings did not always return to their former nest on the first night. In eight out of 17 cases we observed, fledglings were led to other nests in the vicinity. Twice the female gave up her roosting nest; three times we saw the male direct the young to his nest, while he sought sleeping quarters elsewhere. Breeding nests of a previous brood were sometimes used, and even a nest which a juvenile wren had built was used. Antevs (1947: 42) reported that a secondary nest was used by fledglings. The reasons for these variations in nests used for roosting were not always evident. Some were, no doubt, brought on by disturbances, for the wrens moved from obviously hazardous locations into our lot for greater safety. Occasionally at such times, especially when the fledglings became separated, the adults directed them into two or three nests in the vicinity instead of into one nest. On the following evening, all of the young were usually sleeping together. A change to another nest only 2 feet (0.6 m) away from the breeding nest in cholla 17 is unexplainable, as is also a similar switch to a nest 5 feet (1.5 m) away in cholla 5. In both instances the old nests were in perfect condition. A brood from the adjacent territory to the north in 1947 spent its first night in the male's nest, but the young returned to the old breeding nest the second night. Earlier in the same year, a fledgling from the south half of our lot inadvertently strayed into the territory to the north, and spent the night in the roosting nest of an older group of fledglings. The following morning it found its way back to its parents.

In the course of the first 2 to 3 weeks, the adults sometimes shifted the fledglings to other roosting nests in the territory. The fledglings still slept together, with no apparent difficulties from overcrowding. In fact, we never observed any tendency to seek individual quarters up to the time the wrens had learned to go to roost unassisted. Even after their parents had ceased to help in retirement, the fledglings maintained their communal roosting group for a remarkably long time. Losses occurred frequently, so that the brood was seldom intact when the time came to move into individual nests. The three fledglings of the second brood in 1959 did not separate until they were 52 days old; in 1958, the first brood remained together for the same length of time; the second brood remained together 2 days longer; the third brood of five fledglings all roosted together until they were at least 45 days old. This group finally dwindled to two wrens, which began roosting in separate nests at the age of 70 days, 50 days after fledging. The first brood of 1947 required an even longer time, 75 days, for serious antagonism to develop, but the lack of roosting nests in the constricted territory may have been responsible for the prolonged association.

None of the fledglings constructed roosting nests for their own use at this time. The change from communal roosting to single occupancy merely required a move to some unoccupied nest in the territory, such as an old roosting or breeding nest, or a secondary nest which had been built by the male while the female incubated.

The progressive deterioration of the communal roosting nest was evidently not a cause for the adoption of separate quarters. The nest cavity and its entrance, to be sure, expanded from the movements of the crowded occupants; the floor hardened into a rough, solid mass, and holes appeared in the top, sides, and rear of the nest, through which the wrens squirmed their way outward. Nevertheless, they continued to use the nest. Indeed, we have found fledglings huddled together night after night in nests from which the tops had been torn completely away, leaving the floor exposed to the sky.

We have assumed that antagonism among juvenile wrens is responsible for the break-up of communal roosting. The shift to individual nests was usually accomplished gradually at irregular intervals. One or two juvenile wrens moved into other nests; finally all were in separate roosts. Sometimes the change was erratic, as though from indecision. A wren which chose to roost alone for the first time could occasionally be found roosting with its mates again the following night.

The conflicts at roosting time have been difficult to observe, but there seems no doubt that changes to other roosting nests were not always voluntary. The time came when the first juvenile wrens to retire finally objected to further intrusions. The last one to seek shelter received a hostile response from the occupants of the nest. In this case we heard no sounds from the nest; we did not observe any defensive action at the entrance; yet, from the frustrated *buzz* note of the wren outside it was evident that it had been denied permission to enter. After climbing about in the cholla, *buzz*ing repeatedly at its failure to gain entrance, this wren at last flew off in search of another roosting nest. Probably the most aggressive individual retained possession of the nest.

Feeding

At first, both adults shared the feeding duties. Later, when the female moved into a secondary nest for her next brood, the male took charge of the fledglings. However, it was not unusual to find the female assisting the male in feeding their brood. By the time the fledglings' exploratory pecking had led to some recognition of useful food items, the male had begun construction of his next secondary nest. He continued to feed the fledglings when they begged, but they were left unattended for longer periods of time while he worked on his nest.

We have observed exploratory pecking by Cactus Wrens at the age of 21 days, a day after fledging. It probably occurs as soon as the birds are fledged, perhaps even earlier at the nest entrance. At first this pecking appears rather aimless; it is not a deliberate sampling of possible food, for apparently nothing is tasted or picked up. Branches of trees and bushes and cholla joints are gently pecked; vines and grass stems are grasped and pulled and then released. Even nearby fledglings were occasionally pecked! These random probings were noted only when the adults were absent. They indicated a developing interest in the immediate environment.

Self-feeding is established slowly. Some fledglings were observed picking up food at the age of 35 days (15 days after fledging), but begging is frequent at that time. Food is not recognized as such for some time, even when it is provided by a parent. On 20 February 1958, three fledglings, then 29 days old, gathered close to the male as he fed them dog food from a small pan on the lawn. In its impatience to be served, one of the fledglings stepped into the middle of the pan, its feet sinking into the soft mash. It stood there, bill open wide, begging, waiting to be fed. Again, on 4 April 1959, three fledglings, 25 days old, were seen on the ground beneath some creosote bushes. They poked here and there on the ground, occasionally pulling at a Bermuda grass stem trailing on the lawn. A short distance away, a male House Sparrow was eating a small piece of potato which we had supplied. Suddenly H-90 dashed at the sparrow and drove it aside. The sparrow soon returned and again the fledgling dashed at it. This time the wren stepped upon the potato as though to defy any further trespass. Then, when it stepped down, it ignored the food. Sometime later it pecked at the potato, but even though some of it adhered to its bill, the food was not swallowed.

Opportunities for learning to find and recognize food increased rapidly as the fledglings followed their parents on the ground. Often they stood behind or around the male as he searched the ground or lower branches of shrubs for insects. This would appear to be a perfect situation for the acquisition of food gathering habits. Yet, exploratory pecking came first, without parental instruction, and it continued to expand as more energy was acquired. Less time was spent in resting and waiting to be fed. Evidently curiosity led the fledglings to explore and investigate everything in their vicinity. Some, if not all, of their behavior at this time may come from a normal development, as regular and predictable and as stereotyped as the growth of their bodies. Rand (1941: 222) found that fledgling Curve-billed Thrashers he raised in captivity started digging at the age of 19 days. He concluded that "the manner in which these birds sought for, picked up and ate food appeared to be rather rigidly determined innately." Judging from our field experience, we feel the same conclusion could be demonstrated with hand-raised Cactus Wrens.

The transition to self-feeding began on the ground. Fledglings did not forage in trees or bushes at first. Exploratory pecking soon led to the picking up of various irregular objects from the ground. When H-94 was 38 days old, we saw it turning over stones an inch or so (2.5 cm) in length. At this age fledglings peered under flat objects, such as pieces of paper and leaves, by poking their bills beneath the object and lifting it. It is typical adult procedure to push aside or raise up, by means of the bill, dead cholla joints which litter the ground, in order to find insects beneath them. At the age of 41 days, a fledgling pecked at and ate from a dried slice of white bread on the lawn. Then it poked its bill under the slice and raised it to peer beneath it. When the slice suddenly fell flat, the wren jumped into the air in alarm. As with Rand's thrashers, our Cactus Wrens no doubt learned from experience what to eat and what to reject.

The stimulus which elicits begging appears to lie in the attitude of an approaching wren. Food does not have to be present and in view. It is doubtful if the parents are recognized at first; a fledgling will sometimes beg with wide open bill when suddenly confronted by another fledgling. One of a brood, which had just gone to roost, begged from a juvenile wren which came to the entrance of the nest in search of a place to sleep.

Independence is attained at about the age of 50 days, but even 10 to 14 days later the young wren may beg when its parent approaches. It continues to utter the begging note as it feeds itself. The gradual cessation of this innate begging pattern, and its replacement by exploratory pecking, is perhaps partly governed by internal conditions, as Rand has suggested (1941:221). As the frequency of begging diminishes, irregularities in behavior can be observed. Begging may occur, although food is not desired. On 3 March 1958, two fledglings, 40 days old, in company with their male parent, HM-73, were begging occasionally while they searched for food on the lawn. HM-73 fed one of them; then he turned to feed the other. Although this fledgling appeared to be begging as it stretched forward with open bill, it now moved its head to one side, evidently to reject the food which the male extended toward it. Again the male stepped nearer, and again the fledgling, with open bill, turned aside in a quarter circle as before. The male gave up.

The adult wren responds to the begging note and the gaping bill of the fledgling. Sometimes in the strenuous hurry of simultaneously feeding the young and constructing a secondary nest, the begging stimulus produced a startling effect. At the sight of the open mouth, the immediate, automatic reaction of the adult wren was to place something in it. Thus the unsuspecting fledgling received a mouth full of cotton, grasses, or other nest material instead of food. Meanwhile the puzzled parent stood by watching the equally puzzled fledgling chew the unpalatable morsel until it was dropped from its bill. Evidently taste and not sight at this stage determined whether an object was eaten or discarded.

The development of song

Opportunities for learning adult vocalizations are so numerous and so frequent that it is virtually impossible in the field to determine what is innate and what is learned. For nearly 3 weeks the nestling is subjected to the complete adult repertoire. The standard song is repeated over and over; the warning *tek* note, the danger *buzz* note, and the location *rack* note are all uttered so frequently in the vicinity of the nest, that it would seem to be only a question of learning the meaning of each sound. We have heard the *buzz* note from day-old fledglings, perhaps uttered in annoyance or surprise at our close approach. The *tek* note has been heard 2 days after fledging, but this sound is not often used by fledglings.

Lanyon (1960) has defined three phases of song development in passerines, which he terms subsong, rehearsed song, and primary song. The first subsong in the Cactus Wren may begin at the early age of 30 days, only 10 days after fledging. At first the subsong is a weird, up and down, rambling jumble of low, grating, rough notes, a ragged imitation of the adult primary song, and it is short in duration. In a few days it lengthens to two or three times the length of the adult song; then it is interspersed with and often ends with *scri* notes. These *scri* sounds which we have heard so frequently in the subsong are similar to the adult territorial warning and dispute vocalizations. No call notes of fledglings are incorporated in the subsong; the fledgling has only one distinctive call note, the begging *dzip* note. One could describe the subsong as conversational, for the fledgling appears to be talking to itself. The song is not uttered from any elevated point, but is produced while the wren is prowling around in the shrubbery or on the ground. There may be from 6 to 11 songs per minute.

It is difficult to designate any later, intermediate phase of rehearsed song. There is only one adult song pattern. The fledgling subsong has this basic pattern, garbled and distorted to be sure, at the very beginning. The addition of the scratchy *scri* note to the subsong is very suggestive of imitation. At the start of a territorial dispute we found that the vigorous singing of the adults was always followed by sharp *scri* sounds as the excitement increased. In fact, the song and the dispute note actually combined into a single, uninterrupted series. With the lessening or cessation of disputes at the end of the nesting season, the adult song no longer ended on this note. The sequence of change in the subsong in the young

Cactus Wren apparently followed the same course as in the adult. The *scri* notes were gradually dropped, and the rambling, grating sounds were smoothed out into an even, uniform series of adult *rar rar rar rar* syllables. The standard song pattern was, of course, the one most frequently heard by the young wrens. However, there is the possibility that the forceful impressions of a territorial dispute could contribute to the fixing of the additional scratchy note in the subsong.

It seems improbable that this aggressive or defensive scratchy note is an indication of a growing aggressiveness in the young wren. We observed little, if any, territorial aggressiveness in juvenile wrens. True, they became excited and took part in boundary disputes which the adults originated; they unintentionally caused disputes by penetrating into forbidden territory. The subsong began and continued while the family moved about as a unit of friendly, or at least tolerant, individuals; the fledglings slept together each night. In fact, the song appeared to be independent of and unrelated to any of the normal activities. We never observed that the juvenile song elicited any antagonistic response from the parents. Evidently the adults had no difficulty in distinguishing between the subsong of their offspring and the adult song of an adjacent territorial male. The *tirrup* variation of the adult *rack* location call note was often used by the juvenile wrens.

The rapid dispersal or loss of banded juvenile wrens prevented us from determining the sex of the birds which sang. At least 3 or 4 months elapsed before the adult song was attained. By that time most of the young in our vicinity had vanished.

Bathing

Our hundreds of observations of dust bathing by adult Cactus Wrens indicate that this act is a ritual as regular and predictable as their nightly retirement. Each evening, a few minutes before going to the roosting nest, the adult wrens took a dust bath. The dusting place was usually less than 100 feet (30 m) from the nest; in our lot, the place chosen was sometimes a hole used by House Sparrows. From frequent use this hole reached a depth of an inch or so (25 mm) and a diameter of 2 or 3 inches (5 to 7.6 cm). When this was not available, a few preliminary pecks on the bare ground were sufficient to loosen a little dust. Then the wren squatted down and spread its wings slightly, while it ruffled its body feathers and shook its wings until the dust penetrated the feathers. Sometimes the side of the head was rubbed into the dust and the tail was swished from side to side. This done, the wren resumed its slow movement toward its roosting nest. Occasionally it would return and repeat the dust bath before going to roost. In contrast to the House Sparrows, which could be found dust bathing practically any time of day, the Cactus Wrens dusted themselves almost entirely in the evening. We have only a few observations of this activity at other times. Woods (in Bent, 1948: 229) failed to observe dust bathing, very probably, we believe, because he was not present at the roosting time. The act could very easily be missed, for it takes only a few seconds to perform.

There is some evidence that dust bathing may be an innate form of behavior, with learning assisting to perfect it. Fledglings do not ordinarily witness the act until they are able to go to roost without parental direction. This stage is reached in 2 or 3 weeks after fledging. Yet, we once observed a fledgling dust bathing only 6 days after it left the nest. In this instance, the wren pecked the ground and tried unsuccessfully to stir up some dust by squatting down and sliding in a half circle with wings held horizontally. Then it moved into a nearby hole made by a House Sparrow, with better results. This occurred at 1700 in May. Perhaps the fledgling was imitating an adult, but at the time there were no other birds in the vicinity. Other fledglings, 12 to 14 days out of the nest, have also been noted dust bathing just before retiring. Later, when their retirement time began to coincide with that of the adults, the opportunities for learning increased. In late May 1961, three juveniles of the first brood, at least 60 days old, now feeding independently, but still uttering their begging call note, took dust baths in our front yard. The first two merely dipped their heads to the ground and squatted quickly to touch the dust. The third advanced to within an inch (2.5 cm) of its male parent, and watched as the latter thrashed around in a dusting hole. Soon the male moved aside, permitting the waiting wren to step in. In a thoroughly competent fashion the young bird stirred up a cloud of dust, even pecking more sand into the hole while it crouched inside. It was a perfect situation for learning.

On another evening, two fledglings, 18 days out of the nest, stopped at a dusting hole. One of them looked on in apparent surprise as the other stepped in and dusted itself. Perhaps the performance seemed a little aggressive, for the second wren leaned forward in an attempt to peck the first. It was repulsed, but it tried again. This peck was returned as the dust bath ended. We observed no further conflict while the two fledglings moved toward their roosting nest. Perhaps this was an instance of the tardy appearance of this innate behavior pattern in one of the fledglings.

We observed that both adults and immature wrens occasionally bathed in our sunken bird pool, and in the saucers around the ornamental trees. Now and then, on the short grass of our recently sprinkled lawn, we have found them sliding about in an effort to wet their bodies or perhaps to cool off. Opportunities for these activities must be rare in the normal desert habitat.

The family break-up

The family bond is broken gradually by the adults. As the fledgling approaches independence, begging decreases and the parents are confronted with fewer stimuli to feed their young. Begging is finally ignored. It seems like a mutual loosening of the bond between the two. Actually it is the adult which loses interest; another brood will soon be ready for parental supervision.

The juvenile wrens remain in the territory for some time. They may still follow their parents as they search for food, and they participate in territorial disputes, siding with their parents. Pecking of fledglings by adults was rarely observed; when we did record it, it seemed to have little relation to the problem of dispersing the young. In 1946, 19 days after fledging, when the offspring were still roosting in a group, we saw the female approach one of the fledglings on the ground and peck it. The young wren crouched and tried to avoid the sharp blow on its head. There was no further attack.

In the days following fledging, the young responded to the song of the adults. It led them to food and to their evening roosting nest. Later, when they were learning to shift for themselves, the need for a directional song gradually vanished. Its stimulus, however, still had some effect, and sometimes it led to confusing reactions. Once, as we observed three fledglings about to fly to their roosting nest, the adult male sang from the roof of our house. Immediately the three wrens turned and flew to the roof. As each one arrived, it received a vigorous peck. The last fledgling tumbled over the edge and clung upside down by one foot. Then the fledglings regrouped and proceeded to their roost, while the male remained at his station. Occasionally in the course of the day, juvenile wrens would fly to a singing male parent and climb up the post on which he perched. When they reached the top, they were repulsed with a vicious peck which sent them tumbling down.

Chases were noted now and then, but the juvenile birds were reluctant to fly more than a few feet. The most obvious conflicts occurred at roosting nests. Attempts by young wrens to enter roosting nests of adults in the evening were usually thwarted at once. Curious wrens investigating nests under construction were pecked and driven off. Some were ejected from roosting nests. Generally this occurred shortly after a new brood had been fledged. Juvenile wrens carrying nest material seemed to be particularly subject to threatening drives by their parents.

Nest building by juvenile Cactus Wrens

The innate behavior pattern of building roosting nests appears very early in the young Cactus Wren. We once observed a fledgling only 12 days out of its nest toying with a feather. At first, feathers and short grass stems are picked up and then dropped. Later, they are carried to cholla branches or into roosting nests. It is difficult to escape the conclusion that these acts are mere imitations, for the adults at these times were gathering nest materials in full view of the fledglings. Admittedly, much data have been missed here; the roosting behavior of the wrens becomes incredibly involved in the course of the most productive seasons. By the time the first brood approached independence, another brood

Table 14.1. Nest construction by first-year Cactus Wrens.

Wren	Fledged	Carried nest material	Age in days	Began construction of new nest	Age in days
H-25	19 April 1941	23 May 1941	54		
H-33	5 June 1941			15 Aug. 1941	91
H-42	5 May 1944	10 Aug. 1944	117	17 Oct. 1944	185
H-51	5 May 1947	1 July 1947	77	16 July 1947	92
H-52	5 May 1947	19 June 1947	65		
H-74	11 Feb. 1958	22 Feb. 1958	31		
H-78	25 March 1958	3 May 1958	59		
H-81	5 May 1958	14 June 1958	60		
H-82	5 May 1958	14 June 1958	60	22 July 1958	98
H-83	5 May 1958	15 June 1958	61		
H-84	5 May 1958	14 June 1958	60		
H-85	5 May 1958	1 June 1958	47		

arrived to confuse the daily routine, and complicate the evening retirement.

The dates when 12 of our first-year banded Cactus Wrens of known age were first observed with nest materials are given in Table 14.1. A nestling age of 20 days has been added in each instance. These dates may be far from representative, when one considers that over 100 wrens were fledged in our research area from 1939 to 1960. Five of these 12 wrens disappeared less than a month after being seen with nest material; two of these were gone the following day. The 117-day gap in the record of H-42 cannot be accounted for, except by assuming that we failed to watch closely. This wren began carrying lining material in August to a nest which its male parent had constructed in June. H-42 occupied this nest and continued to bring lining material for several weeks. It was not until October, when H-42 was practically finished with the post-juvenal molt, that it constructed a new nest. H-52 finally obtained possession of nest 15C, the male's former roosting nest, by excluding its bedmates. H-52 and two other fledglings had roosted in this nest for some time. For 10 days thereafter it carried lining material to this nest. In August, H-52 suddenly appropriated a nest which the female wren was building; then it began the task of finishing it. Later, it returned to nest 15C, where it remained until we lost it in December. H-78 had a similar erratic history. It worked entirely on old nests. H-33, H-42, and H-82 (Table 14.1) finished their new nests. These nests were indistinguishable in construction from nests built by adult wrens. H-51 disappeared 2 days after starting a new nest. The average age at which nest materials began to be carried by these young wrens was 62.8 days; the average age at which new nests were started was 116.5 days.

Table 14.2. Nest construction by first-year Cactus Wrens of unknown age.

Wren	Date banded	Repaired old nest	Began construction of new nest
H-7	4 Aug. 1939		13 Oct. 1939
H-20	28 June 1940	3 July 1940	25 July 1940
H-21	28 June 1940		20 Oct. 1940
H-22	8 Sept. 1940	30 Sept. 1940	27 Dec. 1940
H-30	18 May 1941	3 Aug. 1941	20 Oct. 1941
H-56	9 Aug. 1952		10 Aug. 1952
H-66	25 July 1954		25 July 1954
H-95	14 Aug. 1960		14 Aug. 1960

In addition to the 12 wrens of known age, we have data on 16 other young wrens that strayed in from adjacent territories and endeavored to remain. Eight of these we succeeded in color banding (Table 14.2). We trapped most of them shortly after they were discovered to be roosting in our area. Eight other juvenile wrens, which we could not trap, started nests from 15 July to 15 September. Five of these nests were new. Mid-July would thus seem to be the earliest that young wrens begin construction of new roosting nests. The availability of good nests is probably an important factor at this time: why build a new home if the old is satisfactory?

Recognition display by juvenile Cactus Wrens

We have observed the adult type of recognition display only twice from juvenile wrens. On 12 August 1956, a young wren *buzz*ed in our front yard as it moved about carrying a grass stem. Suddenly another young bird landed 3 feet (0.9 m) away. It spread its wings in display and growled in typical adult fashion. The first wren appeared startled, but it did not respond to the second wren or leave the yard. Soon both of them were searching for food on the ground. Five days later another display-growl occurred as two juvenile wrens flew to the top of an electric pole. One wren clung to the side of the pole about 2 feet (0.6 m) below the other, but it did not advance toward the other wren. None of these wrens was banded; their fledging dates are unknown. Whatever the date of fledging, August seems to be an unbelievably early date for a young wren to begin to assert itself by means of the adult recognition display.

Juvenile helpers

We had long been aware that "helpers" were a possibility among our Cactus Wrens. Skutch (1935: 269-273) reported that the practice was common in the nesting activities of the southern relative, the Banded-backed Wren (*Campylorhynchus zonatus*). As early as 1947, we noted that young of an earlier brood sometimes permitted a fledgling of the next brood to share the roosting nest. At one time two of the first brood and one of the second brood roosted together for several nights. In these instances, the young fledgling was directed to its nest by its parents. Then, when all was quiet, the older wrens crept in. This was probably only passive tolerance, or an inability to distinguish one from the other. Generally, we observed that juvenile wrens not infrequently pecked the fledglings of the following brood when they met.

The situation on 11 May 1958 was as follows: the female, HF-71, was incubating four eggs in nest 25H; the male, HM-73, had just begun construction of secondary nest 5M. In between work on the nest, he fed fledglings H-81, H-82, H-83, H-84, and H-85 of the third brood, which had left their nest on 5 May. Two of the four wrens of the second brood, H-78 and another, not color-banded, had survived, and remained in the territory. They had been fledged on 25 March.

At 1015, H-78 kept up a constant series of subsongs as it pecked on the ground and turned over small objects. Now and then it chased its sibling of the second brood, and frequently it seemed to be "shadow-boxing," for it snatched at objects on the ground and dashed around, changing direction abruptly. Once it came face to face with H-84, which opened its bill wide as though expecting to be fed. H-78 turned rapidly and ran back into the tangle of bushes.

At 1545, H-78 came with a small insect, landing on the ground below one of the fledglings. When the latter begged, H-78 flew up and fed it. Again H-78 found some food and fed the other. For the third time, H-78, holding a thin green caterpillar, flew up toward H-81. Suddenly one of the wrens in the group began uttering the *buzz* alarm note at a Round-tailed Ground Squirrel standing upright beneath the creosote bush. H-78 approached H-81, but the latter chimed in in the general chorus of alarm. H-78 followed the *buzz*ing H-81 upward. No begging action took place, and H-78 stopped following H-81. Soon it fed another fledgling. Then, when H-78 flew down, two of the fledglings followed, begging. While H-78 searched the ground, finding nothing, HF-71 arrived and fed the two young. The other juvenile wren now appeared and looked on, but it had no food.

Later this same juvenile wren was confronted by a begging fledgling. Receiving no food, the fledgling crouched. The juvenile wren advanced and pecked it on the head. Then it ran to a second fledgling which was not begging and pecked it, too. Finally a third one received the same hostile greeting.

At 1700, H-78 again fed H-81; on the next day, it fed H-82; on 18 May, we saw it feed H-82; on 19 May, it fed H-81 and H-85; on 25 May, it fed H-84. Very probably it had been feeding the entire brood while the male was absent. Our observations were not continuous. On the evening of 19 May, we saw H-78 fly to cholla 5 with one of the fledglings. It watched while the fledgling squeezed into nest 5K, where three others had already retired.

An almost exact parallel juvenile behavior occurred in 1960. H-93 of the first brood had been fledged on 7 April. The next brood of three fledged on 20 May. We saw H-93 feed a fledgling on the evening of 23 May. Then both of the wrens flew to a roosting nest near our west fence. Soon the male arrived and induced two more fledglings to enter. Finally the female poked her head inside, causing H-93 to pop out quickly.

Three days later we saw H-93 pick up a leaf on our back lawn and then drop it. The wren then dashed rapidly in short spurts in various directions. One of the fledglings began following it. Again H-93 picked up objects and dropped them. The fledgling opened its bill and begged, but it was not fed. At 0800 on 28 May, we again saw H-93 feed a fledgling. Half an hour later, we saw H-93 turn around to find a begging fledgling at its heels. The fledgling squatted; H-93 pecked at the ground but apparently found no food. Then it circled the fledgling, and pecked it on the back. The latter crouched, turned to face the other, and continued begging. H-93 again pecked at the ground; then it placed its bill in the open bill of the fledgling. It held it there for a second or two, but no food could be seen. Suddenly H-93 began running erratically, a foot or so at a time, pecking at grass and stones; it even lifted up a tuft of grass and dropped it. Again it came upon the cowering, begging fledgling; again it circled and pecked it somewhat tentatively as though it expected some other response.

This apparently irrelevant behavior suggests that both H-93 and H-78 had encountered a new situation and were unable to cope with it. Here was an unexpected request for food at a time when no food was available. Evidently in a short time they were able to learn to do something which they normally would not do for another year.

Intensive day-long observations of Cactus Wren families would probably reveal more of these interesting helper episodes. H-78 was 67 days old, and H-93 was 66 days old, when we first discovered that they were feeding the fledglings. We do not know when these wrens first began feeding fledglings, but the fledgling of 1958 had been out of its nest only 7 days; the one in 1960 had been out of the nest 4 days. H-78 and H-93 could not have been feeding these fledglings very long. In 1960, at least, we may have witnessed the beginning of the feeding of fledglings by juvenile wrens, with all the frustrating problems and irrelevant solutions.

15. Nesting Success

The hatching and fledging success for the Kleindale Road Cactus Wren nests is presented in Table 15.1. We have regarded a nesting attempt as successful if at least one wren was fledged. Six of the 22 first clutches failed completely. The set of four eggs in 1941 was destroyed; the female wren was missed the same day. HM-23 found a new mate, HF-29, who laid five eggs in their new nest. In 1944, he had another female, HF-39. The first clutch of three eggs failed to hatch. They tried again in a new nest, but the first egg or two were destroyed almost at once. Their third attempt in another nest, this time with four eggs, succeeded. In 1954, the first clutch of four eggs was abandoned when the noband female died in the nest. The noband male obtained another female, who laid only three eggs in the next clutch. This was probably her first set of the year. The first try in 1957 by HM-70 and HF-71 failed when the three eggs and the male were lost. When she found another mate, she laid four eggs in her new nest. Although there is a hint here of a tendency to lay a larger clutch after the failure of a first attempt, in only one of the above instances was the same pair involved. A female was replaced in the first and second examples, and a male in the fourth. An increase in clutch size, after a failure, would appear advantageous to the species, provided the added burden of feeding could be carried successfully. In three of the years, 1942, 1945, and 1958, clutch size increased after a successful first brood, and, with the exception of one of the clutches in 1942, it remained as large as in the first brood in the other years. The estimate of three eggs in 1942 could be low.

Table 15.2 summarizes clutch failures. The failure of a nest in 1944 after one egg was laid has been excluded, and the four nests at which fledging could not be determined have been excluded from the summary of percent fledged. In this table, the rapid decline in number of clutches attempted after the first, and the increase in failures after the second, are

Table 15.1. Hatching and fledging success of Cactus Wren on Kleindale Road.

First clutch

Year	Nest	Eggs laid	Hatched	Percent hatched	Fledged	Percent fledged
1939	28B	3 (est.)			1	
1940	6M	4	4	100	4	100
1941	35B	4	Destroyed by boys			0
1942	7E	4	4	100	4	100
1942	43B	4	4	100	4	100
1943	100	3	3	100	?	
1944	6AB	3	Eggs disappeared			0
1944	66A	3	Eggs disappeared			0
1945	14C	3	1	33.3	1	33.3
1945	37D	4	Nest destroyed			0
1945	51C	3	1	33.3	?	
1947	19C	4	3	75	3	75
1947	6AJ	4	3	75	3	75
1948	6AK	3 (est.)			3	
1952	78A	3 (est.)			1 (est.)	
1953	25C	3	3	100	3	100
1954	25D	4	4	100	0	0
1955	6AQ	3 (est.)			3	
1956	67E	4	?		2	50
1957	93B	3	Eggs disappeared			0
1958	5K	3	3	100	3	100
1959	17L	3	3	100	3	100

Second clutch

Year	Nest	Eggs laid	Hatched	Percent hatched	Fledged	Percent fledged
1939	6G	3	3	100	3	100
1940	32A	4	4	100	4	100
1941	27B	5	5	100	5	100
1942	6T	3 (est.)			3	
1942	60A	5	4	80	?	
1944	6AC	1	Egg disappeared			0
1945	23F	4	4	100	4	100
1945	46E	4	3		?	
1947	6AK	3 (est.)			2 (est.)	
1947	75A	3 (est.)	2 (est.)		0	0

very striking. Although as many as six clutches were laid in a season, the maximum number of broods raised was three. We can summarize as follows: in one of the years, no young were fledged; in four of the years, one brood was raised; in nine of the years, two broods were raised; and in four of the years, three broods were raised.

Our infrequent visits to the Santa Rita Experimental Range unfortunately precluded the evaluation of nesting success in that area.

At Saguaro National Monument, the nestlings received bands when they were 8 to 12 days old. If the nestlings appeared normal and vigorous

Table 15.1 cont.
Second clutch (Continued)

Year	Nest	Eggs laid	Hatched	Percent hatched	Fledged	Percent fledged
1953	56B	3			2 (est.)	
1954	6AR	3	1		0	0
1955	27E	3 (est.)			3	
1956	27H	4			2 (est.)	
1957	96C	4	4		3 (est.)	
1958	25G	4	4	100	4	100
1959	P5	3	3	100	3	100

Third clutch

Year	Nest	Eggs laid	Hatched	Percent hatched	Fledged	Percent fledged
1939	6H	3 (est.)			0	0
1941	35C	5	3	60	3	60
1944	23E	4	4	100	3	75
1945	6AF	4	1 (est.)		0	0
1953	6AP	3	2 (est.)		2 (est.)	
1955	6AT	3 (est.)	1 (est.)		1 (est.)	
1956	17D	4	0		0	0
1957	27I	4			3	75
1958	5L	5	5	100	5	100
1959	P5	3	0			0

Fourth clutch

Year	Nest	Eggs laid	Hatched	Percent hatched	Fledged	Percent fledged
1939	7D	4	4	100	3	75
1941	34A	3 (est.)	0			0
1944	6AC	3 (est.)			3 (est.)	
1956	6AV	4	1 (est.)		0	0
1958	25H	5	3	60	0	0

Fifth clutch

Year	Nest	Eggs laid	Hatched	Percent hatched	Fledged	Percent fledged
1941	32B	4	1	Destroyed by cat		0
1958	17J	4	0			0

Sixth clutch

Year	Nest	Eggs laid	Hatched	Percent hatched	Fledged	Percent fledged
1958	92D	4	0			0

Table 15.2. Summary of Cactus Wren clutch failures, Kleindale Road.

Clutch	Number of clutches	Average eggs per clutch	Clutches that failed	Percent of failure
1st	22	3.41	6	27.3
2nd	16	3.63	3	18.7
3rd	10	3.80	4	40
4th	5	3.80	3	60
5th	2	4.00	2	100
6th	1	4.00	1	100

at that time, and the adults continued to feed them in the course of the following days, we assumed that all survived to fledging. If one or more of the color-banded fledglings could be identified later, the nesting attempt was termed successful. The 6-year nesting success of Cactus Wrens in the Monument research area was 68.8 percent of a total of 154 nests (Table 15.3). The Kleindale Road locality, with 37 successful nests out of a total of 56, attained a 66.1 percent success, a remarkable agreement in hardly comparable samples. Table 15.4 gives percentage success of the first, second, and third clutches. Corresponding figures for Kleindale Road are 72.7, 81.3, and 60 percent. Again the differences are slight. In both localities, successful second clutches outnumbered first and third clutches, and the third was least successful.

The highest percentage of successful clutches, 85.7, occurred in 1967, when nesting began unusually late, after a winter of low rainfall and a very poor growth of annual plants. Sixty percent of the Cactus Wren pairs laid only a single clutch. The lowest percentage of successful clutches, 57.9, occurred in 1968, a year of abundant annual plants.

Early laying afforded sufficient time for ten pairs of wrens to attempt three clutches before the season ended. Losses, however, were unusually heavy in the first and third clutches. In 1966, following the heaviest winter rainfall in the 6 years, six of the nine females laid three clutches.

Table 15.3. Cactus Wren nesting success, Saguaro National Monument.

1963	1964	1965	1966	1967	1968	Total
Clutches						
17	20	35	23	21	38	154
Successful clutches						
14	13	23	16	18	22	106
Percentage successful clutches						
82.4	65.0	65.7	69.6	85.7	57.9	68.8
Eggs laid						
55(15)	61(24)	116(18)	83(27)	62(9)	137(6)	514(99)
Eggs hatched						
45(12)	43(12)	101(18)	75(24)	59(22)	101(6)	424(94)
Percentage eggs hatched						
81.8	70.5	87.1	90.4	95.1	73.7	82.5
Fledged						
42(12)	38(9)	69(9)	57(18)	50(21)	76(6)	332(75)
Percentage eggs fledged						
76.4	62.3	59.5	68.7	80.6	55.5	64.6
Percentage hatch fledged						
93.3	88.4	68.3	76.0	84.7	75.2	78.3

Figures in parentheses are estimated nests with 3 eggs each.

Table 15.4. Percentage of success of first, second, and third clutches, Cactus Wren, Saguaro National Monument.

Clutch	Number of clutches	Successful clutches	Percentage successful clutches
1st	78	54	69.2
2nd	59	42	71.2
3rd	17	10	58.8
Total	154	106	68.8

The average clutch per territory rose to 2.6, with 3.6 eggs per clutch. Losses were light; success averaged 69.6 percent.

In Table 15.5, annual successful clutches show a decline from the first clutch to the second in four of the years, a small increase in 1964, and a much larger increase in 1968. Third clutches show high success in 1966, but very low success in 1968. The outcome of single, double, and triple nestings is shown in Table 15.6.

Fourteen (73.7 percent) of the 19 single clutches fledged young successfully — two in 1963, one in 1964, two in 1965, eight in 1967, and one in 1968. Four of the 14 adult pairs disappeared a short time after their young fledged, two others in September. Most of the remaining eight pairs nested again the following year. Unknown predators robbed five nests, one of eggs in 1964, another of eggs in 1965, and a nest with nestlings in each of the years 1966, 1967, and 1968. After these failures, three adult males with their noband females, and a fourth banded pair, disappeared, leaving four territories vacant. The male in the fifth (1967) territory remained; his noband mate could not be traced with certainty. He nested again in 1968, with a winter-banded female. Adult losses made second clutches impossible in eight (42.1 percent) of the 19 territories.

Table 15.5
Annual Successful Cactus Wren clutches, Saguaro National Monument.
First number indicates total clutches; second number, successful clutches.

Clutch	1963	1964	1965	1966	1967	1968	Total
1	9-8	11-7	19-13	9-6	15-13	15- 7	78-54
2	7-5	9-6	16-10	8-5	6- 5	13-11	59-42
3	1-1			6-5		10- 4	17-10
							154-106
			Percentages				
1	88.9	63.6	68.4	66.7		86.7	46.7
2	71.4	66.7	62.5	62.5		83.3	84.6
3	100.0				83.3		40.0

Table 15.6. Success of single, double, and triple clutches of Cactus Wren, Saguaró National Monument.
First number indicates number of clutches in each category.
S indicates successful clutch, F failure, in time sequence.

Single clutch	Percentage	Double clutch	Percentage	Triple clutch	Percentage
19-14S	73.7	22(S-S)	100.0	2(S-S-S)	100.0
		9(S-F)	50.0	4(S-S-F)	66.7
		8(F-S)	50.0	3(S-F-S)	66.7
		3(F-F)	0.0	3(F-S-S)	66.7
				2(F-F-S)	33.3
				3(F-S-F)	33.3

Only three of the 42 double clutches failed completely (Table 15.6). Tabulated singly, these 84 clutches had a success of 72.6 percent, surprisingly close to that of the 19 single broods. The pairs which attempted to raise two broods in a season did no better, on the average, than those which attempted only one. The 17 triple clutches did not fare as well. Although there were no complete failures, only two attempts were entirely successful. Tabulated singly, the 51 clutches had a success of 60.8 percent.

Table 15.7 lists breeding nest losses. Despite their greater height, making them less accessible to snakes, nests in saguaros suffered a loss of about 46 percent of the total of all 48 nests lost in both saguaros and chollas. Cholla nests had a loss of 54 percent of the total. In total nests, 22 (40 percent) of the 55 nests in saguaros failed; 26 (28.3 percent) of the 92 in chollas failed. Building nests in saguaros does not seem to be advantageous. Some of the estimated seven nests lost with eggs may have had recently hatched nestlings. Fifteen nests with banded nestlings are included in the total of 30 which were robbed. A few of the losses could have occurred on the evening of the day of fledging, a particularly dangerous time. One egg failed to hatch in each of 18 nests; two eggs failed in each of three nests.

Table 15.7. Cactus Wren breeding nest losses, Saguaro National Monument.

	Nests placed in				
Cause	Saguaro	Saguaro stump	Cholla	Total	Percent
Eggs robbed	7 (est.)		7	14	29.2
Nestlings robbed	11	2	17	30	62.5
Nests deserted	1		1	2	4.2
Nests fell down	1		1	2	4.2
Total	20	2	26	48	
Percentage	41.7	4.2	54.2		

The number of young fledged per pair was 4.7 in 1963, 3.5 in 1964, 3.6 in 1965, 6.3 in 1966, 3.3 in 1967, and 5.1 in 1968. The 6-year average was 4.3 young per pair. Had it been possible at all nests to determine by actual count, instead of by estimate, the number of young which fledged, the Cactus Wren nesting success would probably be somewhat lower. The overall success of 68.8 percent (Table 15.3) agrees well with the 65 percent average in the summary by Nice (1937: 143-144, 150) of the success of various passerine, hole-nesting species of birds. The pouch-shaped nest of the Cactus Wren may not be strictly equivalent to a tree hole nest. Nests in cholla cacti were always accessible to climbing snakes; those in saguaro crotches could seldom be reached unless nearby branches of palo verde trees afforded a bridge. Nests excavated in trees by other birds faced comparable hazards, dependent upon the height of the nest above the ground.

16. The Physical Environment and Cactus Wren Survival

Physical environment

From 1939 to 1961, the temperature maxima recorded at the University of Arizona weather station, about 3 miles (4.8 km) southwest of our home in Tucson, ranged from 110F (43.3C) to 115F (46.1C). The extreme minimum was 16F (−8.9C), in January, 1949. These records may vary considerably, especially in winter, from those in the Kleindale Road study area. At night, cold air from the upper elevations of the adjacent Santa Catalina Mountains moves down into the Rillito Valley trough, often producing frost in our neighborhood, while the slightly higher portions of the city escape. These extremes in temperature must be well within the limits of tolerance of the Cactus Wren. In other localities of their extensive range, they maintain themselves under even higher and lower temperatures.

The microclimate of the Cactus Wren's environment is markedly different from the picture given one by standard weather information. Only when the wrens perched at a 4-foot (1.2 m) height in shade were they subjected to the official temperature. On clear mornings, in January, the ground surface temperature at sunrise might be several degrees lower than the standard air temperature; by 1500, when the January air temperature had reached 84F (28.9C), the ground temperature had risen to 101F (38.3C). More striking variations occur in midsummer. At an air temperature of 106F (41.1C), we have observed a ground temperature of 146F (63.3C) in the sun. We measured the temperature with a Weston metallic thermometer, its stainless steel tube placed horizontally upon the surface of the sandy ground. The errors introduced by reflection from the bright metal tube, and its exposure to air circulation, are on the negative side. Undoubtedly the actual temperature was higher. Shreve (1951: 14) reported rock surface temperatures in the Sonoran Desert of from 150 to 160F (65.5 to 71C).

The Cactus Wrens obtained relief from the midday summer sun by seeking the shade of bushes and trees. In their search for food on the ground, they visited open spaces for only short periods of time, or avoided them altogether. We once frightened a family of wrens from the shade of our house; they returned as soon as we were out of sight. In summer, wrens frequently move slowly on the shady ground or in the lower branches of a mesquite tree, holding their bills open a quarter of an inch (6 mm), as though panting, while they peer into the tangle of shrubbery for food. At the same time they lift their wings slightly to air their bodies. The extensive field and experimental observations on Cactus Wrens under high summer ambient temperatures by Ricklefs and Hainsworth (1968a: 227-233) confirm that the wrens not only tend to forage in shady areas at this time, but also decrease their foraging activities.

The use of a covered breeding nest, which obstructs the direct rays of the sun, is of considerable advantage to the Cactus Wrens. If the roof of the nest completely shades the nest cavity, and if the walls are sufficiently porous to permit air circulation, the temperature of the nest interior approaches the standard shade temperature. At two breeding nests, checked on 29 June and 12 July, we found the shade temperature and the interior nest temperature to be identical, 105F (40.5C) in the first, and 106F (41.1C) in the second. Another nest, with a thinly latticed roof, which the sun was able to penetrate, was checked on 25 May, and found to be 112F (44.4C) inside, when the shade temperature was 107F (41.6C). The inside of a nest of dense construction would probably reach a still higher temperature, the same temperature as the solid wall mass.

There is a fascinating problem in physiology here which is yet to be thoroughly studied. Other desert birds, notably Curve-billed Thrashers, Mourning Doves (*Zenaidura macroura*), and House Finches, build open nests, and often incubate their eggs while sitting under direct sunlight. Mourning Doves do not change places in the heat of the day, but Curve-billed Thrashers alternate in incubation. The female Cactus Wren leaves her eggs periodically to obtain food for herself. We do not know if she can withstand direct sunlight upon her back during incubation, but, in any event, her eggs could probably not be left uncovered for 10 or 15 minutes in the midday sun of June or July without suffering injury.

Some apparent discomfort has been observed at nests on hot summer evenings. Fledglings sometimes roost in the vestibule of their nest, their bills pointing outward, instead of crowding into the interior. Occasionally adults, evidently hesitant, will stand or sit at the entrance of the nest for some time after dark before moving inside.

One would expect water requirements of wrens to be high under these conditions. In the normal desert habitat, pools of water are seldom available for any useful length of time. On our lot we provided a steady supply of water in a bird bath saucer, sunk in the lawn. Drinking by adult wrens from this pool became noticeable in September, increasing to a high in December and January. We have very few records of drinking by adult wrens in July and August. Evidently the insect food obtainable in the winter months does not have sufficiently high water content to satisfy the wrens' needs. With the coming of spring, the rainfall usually dropped to zero, but the winter annual plants bloomed in more or less profusion, supporting fresh succulent insects. As the days grew warmer, the insect population increased, and the wrens seldom came to drink. Strangely, in August we have numerous observations of drinking by immature wrens.

Very little data is available on the quantities or kinds of food items utilized by Cactus Wrens in Arizona. In California, Beal (1907: 64-65) has reported an examination of 41 Cactus Wren stomachs taken in the region from Los Angeles to San Bernardino, in an area near orchards and grainfields, from July to January, inclusive. The results are not separated by months. Animal matter amounts to about 83 percent and vegetable matter, 17 percent. Of the former, beetles and Hymenoptera each made up about 27 percent of the total. There were many ants and a few wasps. Grasshoppers constituted about 15 per cent. Hemiptera (bugs) and Lepidoptera (caterpillars and allies) each made up 5 percent, the remaining 3 percent consisting of a few spiders and unidentified species. It seems obvious that grasshoppers, caterpillars, and ants would generally be unavailable during the colder winter months. Vegetable matter consisted of fruit pulp and weed seeds. Seeds of *Erodium* and *Amsinckia* (both present also in the Tucson region) were identified. We suspect that vegetable matter (dry weed seeds) must form a considerable part of the winter diet in our area.

The time at which a Cactus Wren retires to its roosting nest is evidently governed by light intensity. As day length increased, the wrens followed sunset closely. To some extent, weather conditions, such as cloud cover and rainfall, influenced the time. We have records in March of wrens going to roost on cloudy evenings as early as 8 minutes before sunset. Rain at roosting time may also induce early roosting. Temperature seems to have little effect. Generally, from December to April, both sexes are in their nests by 5 minutes after sunset. We have some evidence that females retire earlier than their mates in the winter and spring; later in the year, we find no constant difference between the two. While incubation was in progress, little change in roosting time could be observed. When feeding of nestlings began, retirement occurred later, sometimes from 11 to 16 minutes after sunset.

The observed time at which the sun was entirely below the horizon in the Kleindale Road study area seemed to be from 4 to 11 minutes earlier than the official astronomical sunset. This was because of the ragged, irregular profile of the Tucson Mountains at the western horizon. As the sun moved northward, sunset followed the saw-tooth outline of the mountains. For a short time we recorded the light intensities in the evening by means of a General Electric photographic exposure meter. This meter was calibrated in foot candles, its scale reading from 0 to 75. (Later, when we obtained a slip-on multiplier, its range could be increased to 750 foot candles.) We realize that considerable error could be introduced in our measurements, for the locations of the light meter varied from time to time, the sky was not always uniformly bright and clear, and too few readings were taken for them to be truly representative of daily, seasonal, and climatic variations. Nevertheless, certain patterns are evident. Sixteen readings were recorded of the roosting time of a male Cactus Wren, from 26 December to 25 March. On seven of these evenings, the pointer of the meter went off scale at 75 foot candles. From subsequent observations we estimate the true values to be from 120 to 140 foot candles. The remaining nine readings ranged from 34 to 70 foot candles; the average was 54 foot candles. This would indicate that roosting usually occurred very close to the standard sunset time, or a little before. Actually the wrens retired shortly after they observed that the sun in our neighborhood had sunk below the horizon. Movement toward the roosting area usually began before sunset, at a fairly high value of light intensity. Cactus Wrens that roosted on our lot, or the adjacent lot to the west, often approached their nests in the evening from the east or northeast. They moved toward the sun, facing it.

We have defined awakening time as the time at which the wren leaves its nest in the morning. Singing by the male usually occurs immediately thereafter. Our few records of awakening have been combined with the more numerous records of first morning song, for a total of 27, in the first 4 months of the year. Male wrens left their nests and sang, on the average, 29.6 minutes before they saw the sun rise. Incubating females were tardier, sometimes remaining in their nest until just before sunrise. As with sunset, the variation in the time of sunrise followed the slope of the mountains on the horizon. The standard sunrise occurred about 10 minutes before the sun appeared over the Rincon Mountains. Only 8 foot candle readings are available in the foregoing tabulations. They range from a (estimated) low of 0.5 to a high of 7, with an average of 2.6. Apparently the Cactus Wren retires when the light intensity is 20 times as great as it is in the morning awakening period. Fatigue may be a factor contributing to early retirement, but it is difficult to prove. At night in the nest, the bird's eyes may grow accustomed to darkness, permitting profitable activity at a lower light intensity in the twilight before sunrise. Not to be ignored in this discussion are the songs of the other desert birds. Curve-billed Thrashers and House Finches habitually began singing earlier than the Cactus Wrens. If there were no sounds to awaken the wrens, hunger of course would eventually bring them forth in search of food.

Windstorms occasionally blew down poorly placed or flimsily constructed nests in the cane cholla. Heavy, prolonged rains waterlogged the older nests until they slipped from their supports and fell to the ground. Jumping chollas, far more spiny and intricately branched, held nests tightly and securely. Heavy rains also thoroughly soaked the Cactus Wrens, turning them into bedraggled tramps. Wrens seldom realized that they could find dry shelter under eaves and patio roofs, or even in their own roosting nests.

Mean precipitation (Weather Bureau, 1965) in the Tucson area from 1930 to 1965 is 11.17 inches (283 mm); it varied from a low of 5.34 inches (135 mm) in 1953 to a high of 17.99 inches (456 mm) in 1964. From August to December 1962, in Saguaro National Monument, precipitation was 6.54 inches (166 mm); in 1963 it was 11.20 inches (284 mm); in 1964 it was 11.53 inches (292 mm); in 1965 it was 15.12 inches (384 mm); in 1966 it was 12.51 inches (317 mm); in 1967 it was 13.65 inches (346 mm); and in 1968, from January to August, it was 8.89 inches (225 mm). These figures are from the headquarters station at Saguaro National Monument.

The temperature means of the University of Arizona weather station could be used without serious error in the research plot in Saguaro National Monument. They paralleled those in the Monument and were slightly higher; the maxima and minima were also higher.

Survival

We banded three-fourths of our adult Cactus Wrens in the period from late September to March. How many, if any, of these were birds which had been fledged here the preceding spring is not known. It was usually safe to assume that if a banded wren had not been seen in the area for a month or so, it would never be seen again. Returns after such an absence were negligible. As a rule, our Cactus Wren population appeared to maintain itself by additions from other territories. The length of time that banded adult wrens remained in our area is shown in Fig. 16.1. The age given is a minimum; some wrens may have been at least a year old when first banded. Others may have been birds of the year, just over their postjuvenal molt. HM-73, our patriarch, banded on 7 July 1957, was last seen on 3 June 1962. He was then at least 5 years old. HM-23, runner-up, was banded on 19 January 1941. He was last seen on 24 May 1945. The average age of the seven banded adult males is 754 days; the average age of the sixteen females is 500 days.

We know that many of our wrens were killed by cats. Some apparently died from natural causes. We found four adult wrens dead in their nests. Three of these were in roosting nests; the fourth was a female with eggs just hatching. Such cases may not be uncommon, for we have several similar records in other localities. Six other wrens were found dead on the ground beneath their nests. Four were adults, and two were immature wrens. In two of these instances, the nests were damaged, but there is no direct evidence that thrashers killed the wrens. It seems more probable that the nests were attacked after the thrasher discovered that they were not defended. Three other wrens were found dead in the yard, the cause of death unknown.

Three immature Cactus Wrens, two females and a male, of unknown origin succeeded in maintaining themselves in our study area. HF-22,

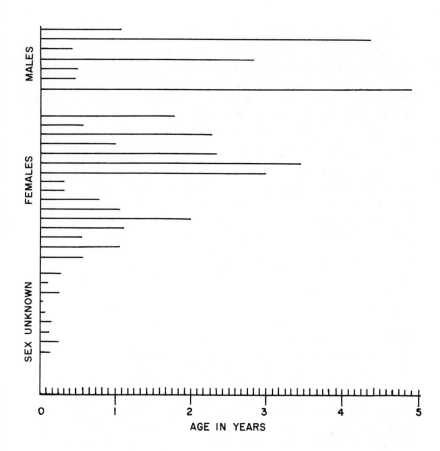

Fig. 16.1.   Length of time that adult Cactus Wrens, banded from 1939 to 1962, remained in the Kleindale Road, Tucson, Arizona, study area.

banded 8 September 1940, was molting when banded, and her age is somewhat in doubt. However, we believe this wren was probably the same immature wren which had been in the territory for some time before. On 27 February 1941, we found her incubating three eggs in nest 35B. She was last seen on 3 March; on the following day we discovered that her nest had been destroyed by boys. HF-30, banded on 18 May 1941, laid her first egg in nest 7E on 7 March 1942. Her four nestlings were fledged successfully on 17 April. In May she was incubating her second clutch in nest 6T; the young were fledged in June. She was last seen on 11 October 1942. HM-66, an immature male, banded 25 July 1954, was last seen on 16 July 1956, after having stayed for two breeding seasons. A nestling, HM-42, banded on 23 April 1944, fledged on 4 May. This male was the only nestling out of 55 banded ones, from 1939 to 1961, who remained until the following breeding season. He was last seen on 20 January 1946.

A small number of the breeding nests were inaccessible; they could not be examined without destroying a large portion of the chollas in which they were placed. Others could not be visited at the proper time. Consequently all nestlings in our vicinity did not receive bands. It is just possible that the percentage of fledglings which survived to breed in our area would be slightly greater if one knew the origin of those wrens which we did not succeed in banding until late autumn or winter.

Survival of banded nestlings, after fledging, is shown in Fig. 16.2. To obtain the actual age, the average age of 20 nestling days should be added to the lines in the chart. Ten of the 55 fledglings could not be found after the first day out of their nests; in 45 days, 41 of the 55 had disappeared. Scattered piles of feathers near fences gave evidence that house cats had been responsible for many of these losses. About half of the 55 fledglings had probably attained a fair degree of independence in their search for food and may have dispersed into surrounding territory. Their fate is unknown; we could not explore the neighboring residential blocks with any success. However, it is doubtful if many of these fledglings survived. Spare roosting nests were not available in the immediate vicinity, and in most of the residential lots there were no cacti in which to build nests. Furthermore, the fledglings had not yet reached the nest-building age. Four of our banded nestlings, after reaching an average age of 7 months, disappeared at the end of December.

The breeding season of 1963 in Saguaro National Monument began with 50 percent of the Cactus Wren population banded. The percentage dropped slightly in 1964; then it rose gradually in the course of the following 4 years, from additions of newly banded winter birds and banded nestlings of previous years, to 87.5 percent in 1968 (Table 16.1). The percentage of banded males remained high in all years, reaching 100 percent total in 1968, but the number of banded females, although showing an increase, was always lower than that of the males. Several factors could be responsible. The sex of the 40 winter-banded adults

Table 16.1. Percentage of Cactus Wren breeding population banded in Saguaro National Monument.

Year	Territories	Banded Males	Females	Not banded	Total	Banded	Percent Males	Females
1963	9	7	2	9	18	50.0	77.8	22.2
1964	11	9	1	12	22	45.5	81.8	9.1
1965	20	18	3	18	39	53.8	90.0	15.8
1966	9	8	4	6	18	66.7	88.9	44.4
1967	16	14	8	10	32	68.8	87.5	50.0
1968	16	16	12	4	32	87.5	100.0	75.0

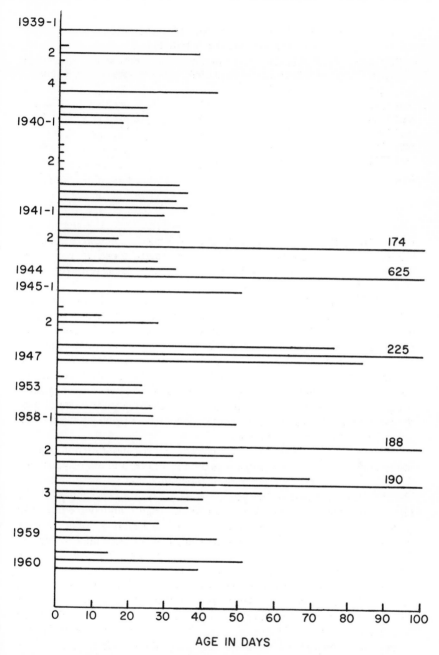

Fig. 16.2.   Survival in days after fledging of banded Cactus Wren nestlings, Kleindale Road, Tucson, Arizona, study area. Off-scale values are indicated by numerals at right. Numerals at left side, 1, 2, 3, 4, indicate broods.

Table 16.2. Loss of winter-banded Cactus Wrens in Saguaro National Monument.

Banding period, winter	Number banded	Number nested	Number lost	Percent lost
1962-1963	19	10	9	47.4
1963-1964	5	1	4	80.0
1964-1965	4	4	0	0.0
1965-1966	17	5	12	70.6
1966-1967	12	10	2	16.7
1967-1968	24	11	13	54.2
Total	81	41	40	49.4

(Table 16.2) that disappeared is unknown. Some females were certainly present in this loss of 49.4 percent. Females may be more wary of traps; they may have a lower survival rate; or they may disperse farther than males.

The number of good, usuable nests in a given area is a rough indication of the number of Cactus Wrens present. It can be used as a measurement of survival, for a sedentary species has no room for its surplus population; only the survivors remain. The course of the monthly counts of nests from August 1962, to August 1968, is shown in Fig. 16.3. These totals include roosting, secondary, and breeding nests. Since breeding nests become multiple roosting nests for a while after the young have fledged, the count of nests at this time is a rather inaccurate measure of the population. There are more wrens than nests. Secondary nests often serve for second or third clutches.

The average number of usable nests found per month from January to December, from 1963 to 1967, was 39, 62, 75, 48, and 69, respectively. The average per territory per month, assuming that all territories were established by the first of the year, was 4.3, 5.6, 3.8, 5.3, and 4.3. The highest numbers of nests were reached in August and September, an indication that the surviving juvenile wrens had begun building roosting nests of their own. The lowest numbers occur from December to March. Here the 4-month average nests per territory become 2.6, 4.3, 3.0, 4.6, and 2.7. Allowing for inaccuracies in determining the condition and occupancy of nests, we can assume that at the low point only a pair of wrens survived in each territory. The all-time low of 16 nests on 1 October 1962, implies that part of the 1963 population must have been composed of immigrants from neighboring areas. The carry-over in the remaining years was sufficient to maintain the population without the aid of immigration.

It was impossible to obtain, in the short space of 6 years, a satisfactory record of the survival of individual adult and young Cactus Wrens, for too many of them were still present in August 1968, when the study

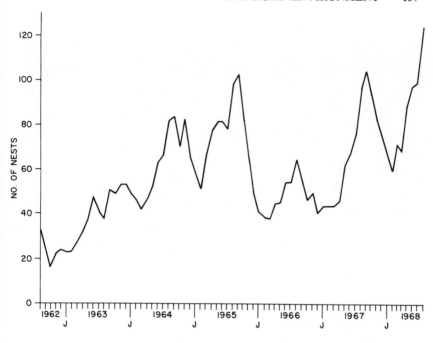

Fig. 16.3.   Monthly count of Cactus Wren nests, Saguaro National Monument, east of Tucson, Arizona, including roosting, breeding, and secondary nests, from August 1962 to August 1968.

ended. Table 16.2 lists the losses of winter-banded wrens. The age of these wrens is unknown; some were doubtless year-old birds, others older. Table 16.3 lists the number of banded wrens, including those in the adjacent extralimital territories, with the number of years they nested. Their breeding numbers decrease rapidly after 2 years. By deducting the wrens present in 1968 from the year-old categories, we find that ten 1-year-old wrens disappeared after one nesting season, four after two, and two

Table 16.3. Cactus Wren survival in Saguaro National Monument.
Figures in parentheses (included in totals) indicate number of banded adult wrens still present in 1968.

Nested, years	Winter-banded males	Winter-banded females	Year-old males	Year-old females	Totals
1	8(4)	10(7)	15(7)	3(1)	36(19)
2	9(6)	5(4)	8(4)	1(1)	23(15)
3	5(1)	2(1)	2	1(1)	10(3)
4	1	1(1)	2(2)	1(1)	5(4)
Totals	23(11)	18(13)	27(13)	6(4)	74(41)

after three. Three 4-year-old wrens were still present in 1968. CF-86, banded 7 January 1965, nesting in 1968, was probably the oldest wren.

Losses and replacements of banded adult Cactus Wrens in the course of the breeding season were few. Changes among the nobands could not be detected, but they must have been rare, for interruptions in nesting activities seldom occurred. In 1965, CM-21 disappeared after his first brood fledged. His noband female, or possibly a newcomer, attracted a noband male for the second clutch in territory VIII. After the failure of their first nesting attempt in territory IV, in 1966, CM-168 and his noband mate could not be found. The pair in territory V expanded their range to include both territories. CM-68 in 1968 found a noband replacement quickly when his mate, CF-174, disappeared after the loss of their first clutch of eggs. The confusing behavior of CF-248 is detailed under "Dispersal."

Unfortunately the presence of large numbers of nobands makes the tabulation of the proportions of first-year wrens in the population an unsatisfactory estimate. A possible minimum, however, may be of some value; it can be obtained by assuming that all the unknowns are adults. The minimum percentages are as follows: 13.6 in 1964, 20.0 in 1965, 11.1 in 1966, 21.9 in 1967, and 11.8 in 1968. An opposite assumption, that all the unknowns are year-old wrens, results in maxima which are obviously too high. The true value must be somewhere in between. For example, the increase from 11 pairs of wrens in 1964 to 19 pairs in 1965 must have come in large part, if not wholly, from year-old birds. There is no evidence yet that adult male wrens leave their territories to set up new ones elsewhere; neither do adult females, in any significant number, abandon their territories unless their mates are lost and no replacement appears.

Using the same set of assumptions, the proportion of year-old male wrens in the male population can be estimated, with the following results, revealing a closer range of values: in percent, 1964, 27.3-54.5; 1965, 38.1-61.9; 1966, 11.1-33.3; 1967, 31.3-68.8; and 1968, 25.0-37.5. The lowest proportions of year-old males occurred in 1966 and 1968, the years when the average clutches per territory were the highest, and the number of triple clutches reached two-thirds of the total in each of those years. It seems reasonable to assume that females were present in the same adult-to-year-old ratio, but proof is lacking. There was one known adult female and one year-old female in 1966. The ages of the other seven are unknown. In 1968, seven of the 18 females were known to be adults; 11 were in doubt.

17. Replacement and Dispersal

We have two chronological sequences of approximately 5 years each, 1941 to 1945, and 1957 to 1961, inclusive, of mated pairs of Cactus Wrens (Fig. 17.1) on the Kleindale Road research area. In these years, the males survived considerably longer than their mates. Data on other years are incomplete. HM-48, banded 11 August 1946, was last seen on 4 June 1949. HF-49, his mate, banded on 28 December 1946, was last seen on 27 May 1950.

In those years when a single pair of Cactus Wrens occupied the entire 10-acre (4 ha) block in the vicinity of our home, the loss of a male in the course of the breeding season was not often immediately apparent. One had to be alert to observe the vacancy; if we did not watch for colored bands each day, the event was missed, for a new unmated male usually moved into the territory at once. The cessation of song appeared to signal that space was available. Our first indication that something had gone wrong was the discovery that a noband wren was singing instead of a familiar banded one.

We experienced less difficulty in following the changes when crowding constricted the territories into smaller units. Upon the absence of a rebuttal territorial song, the adjacent pair of wrens quickly invaded and occupied the vacant territory. When this happened in late February 1948, the widowed female retreated, probably to the edge of the territory, and returned in the evening to slip quietly into her roosting nest. A month later, she vanished; then, after an absence of about 22 months, she reappeared with a new mate and attempted to build a breeding nest. Before the nest was completed, she vanished again.

In April 1940, HF-2 was incubating four eggs when her mate died. The new mate, who arrived at once, had little to do until the nestlings required food; then he assisted in feeding the brood. Once we saw him feed his mate. After the loss of the fledglings, the pair moved eastward, out of the territory.

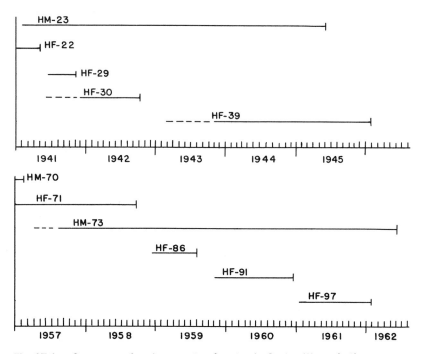

Fig. 17.1. Sequence of replacements of mates in Cactus Wrens in the
Kleindale Road, Tucson, Arizona, study area. Heavy line indicates
banded wren was present; dashed line, present but not banded;
vertical line, last seen.

Females were not always readily replaced, especially after the beginning of the breeding season. On 21 March 1954, when the female had died in her nest, the male began singing vigorously and frequently. Day after day he continued; once he even sang in flight. Six days later he raided a nest box occupied by House Sparrows. At one singing station, on 28 March, at the far corner of the territory, we counted 166 songs in 25 minutes. These were uttered without interruption. In the interval from 0820 to 0926 he sang 274 songs. Finally, on 4 April, a new female appeared. The male attempted copulation the same day. Singing decreased noticeably thereafter, but the territorial aggressiveness did not stop. When we examined the sparrow box on 6 April, two of the newly laid sparrow eggs were missing.

The exact time that replacements occurred in the course of the early winter months could not be determined. The pair bond now became too inconspicuous, and the presence of other wrens in the territory contributed to the difficulty of observing such events.

We have no evidence to support the theory that dispersal of first-year birds is of a genetic nature (Johnston, 1961). The problem in the Cactus

Wren appears complicated by the territoriality of this extremely sedentary species. Once established, the adult wren remains in its chosen territory for life; it mates for life. Dispersal of the immature birds must occur, either voluntarily or involuntarily. There is simply no space available for the young Cactus Wren in this situation where the aggressive behavior of the dominant adult pair excludes all others of its own species. It seems illogical to attribute to the immature Cactus Wren an hereditary tendency to dispersal when its dispersal may be entirely involuntary. Immature wrens assist their parents in boundary disputes in the summer and autumn, but eventually they leave. No data are available to indicate whether their dispersal distances follow a normal probability curve, or any other type of curve. Dispersal in the autumn usually implies the abandonment of a completed, comfortable roosting nest. Such a serious dislocation of routine could only be caused by force, or by repeated disturbances at roosting time. Dispersal immediately upon the attainment of independence must certainly occur in a period of familial instability. The weaning process imposes a strain. A sufficient number of immature wrens have wandered into our neighborhood in the spring to establish the fact that early dispersal does occur. Sometimes these wrens found roosting nests and remained a while, but as long as an adult exercised control, their stay was temporary. Most of our data so far in this restricted area indicate that immature wrens endeavor to remain in the place of their origin, but few succeed.

Faithfulness to a definite territory was well exemplified in the Cactus Wren adult males in Saguaro National Monument. Eleven of 23 winter-banded males nested the following spring in the territories in which they had been trapped; the remaining 12 moved into adjacent areas, establishing new territories, or occupying old ones found vacant because the owners had disappeared. Since trapping did not begin until the winter of 1962, the origin of eight banded male and two banded female Cactus Wrens who survived to occupy territories in the research area in the spring of 1963 is left in some doubt. After the autumn molt, immature birds could not be distinguished from adults. Five males, CM-2, 16, 9, 17, and 11, trapped in territories V, VI, VII, VIII, (recently established), and IX, respectively, remained to breed there in 1963. CM-7, trapped in IX, nested 300 feet (91 m) north, in the extralimital ex1-12. Some of those can be assumed to be former residents, either adults or fledglings of these territories. CM-6, trapped in III, moved northwestward to take over adjacent II; CM-12, trapped in IV, moved northward into adjacent, previously vacant, I. Both may have hatched respectively in territories III and IV in 1962, and were forced now to leave home areas because of intraspecific hostility. Two nobands held IV; CF-13, trapped in VI, moved northward to nest in III with a noband male. CF-10 nested in VII where it had been trapped with CM-9.

There are many assumptions here which seemingly border on mere conjecture, but the repetition of the Cactus Wren's consistent dispersal pattern is so frequent that the pattern must be considered a strong probability. Whenever possible, the male Cactus Wren remained in the immediate vicinity of the place in which he had been hatched; he dispersed only so far as he was forced to go.

Fifteen winter-banded adult male Cactus Wrens survived to breed for two or more years in the Saguaro Monument research area and its surrounding extralimital area. Six of these nested the following spring in the territories where they had been trapped; nine established themselves in adjacent territories without having to cross already occupied land. Seven of the 15 males occupied their territories for 2 years, four for 3 years, and one for 4 years. Three (20 percent) changed territories: CM-83, 245, and 250. CM-83, trapped in the western part of III, possibly raised in IV, moved northeastward in 1965 to establish a new, small territory XV; III was occupied by another male. He nested again in XV in 1966, although III was now vacant. No doubt there were some explorations into the latter region, for in 1967 he assumed control of territory III. CM-245, trapped in ex1-2, set up territory ex1-10, a short distance to the south, but failed that year to attract a mate. In 1968 he moved into territory VIII, whose 1967 male had disappeared. To reach VIII, CM-245 had to cross ex1-2, an active territory in both 1967 and 1968. CM-250 nested in IX where he had been trapped early in 1967. The following year he moved north into ex1-16, leaving his original territory in possession of another male. After his second brood he returned to IX, finding space in the western part of the territory for another brood. The cause of these movements is not entirely clear, but a change of mates occurred.

Adult female Cactus Wrens exhibited a considerably greater tendency to disperse; so much so, in fact, that any conjecture in regard to their origin is hazardous. Thirteen of the 18 winter-banded females were probably already in possession of their territories when they were banded, for they remained there to nest in the spring. The other five settled in adjacent territories. Only eight of the 18 winter-banded adult females survived to breed for 2 or more years. Two of these females occupied their territories for 2 years, and one for 3 years. The remaining five (62.5 percent) changed their breeding territories. CF-86, trapped in January 1965, in what later became territory XVI, disappeared for 2 years, and reappeared in 1967 to nest in ex1-11, 500 feet (152.4 m) to the northeast (Fig. 17.2). She was trapped twice in January 1968, at the north fence in territory I. A month later she was with a new mate in territory VIII, 1,400 feet (426.7 m) south, having traversed the entire length of the research area, passing over at least three active territories, and leaving a year-old banded male from territory I and a noband female in control over her old home range.

CF-174, trapped in XIII, nested there in 1966; she may have been the noband female in that territory during the preceding year. The adjacent, vacant territory VII was probably utilized to some extent. She

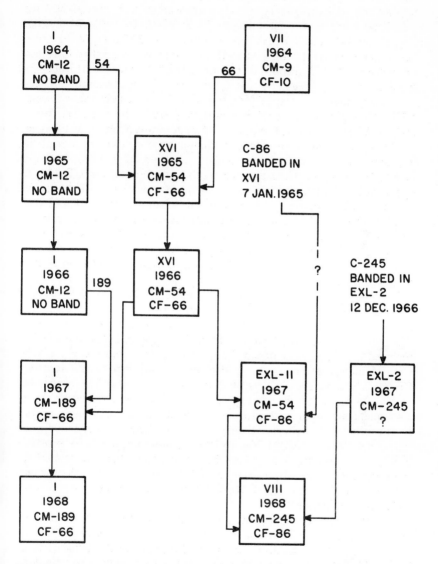

Fig. 17.2. Cactus Wren, Saguaro National Monument. Nesting sequence of CM-54 and CF-66. Squares with Roman numerals indicate each season's territory; CF-66 paired twice with offspring of CM-12, once in 1965, and again in 1967.

nested in VII in 1967 and 1968. Territory XIII, in possession of a new pair, became largely extralimital (now numbered ex1-13) as it moved beyond the south boundary. CF-239, trapped in VII, moved 300 feet (91.4 m) southwest to nest in ex1-13 in 1967. In 1968, a new pair occupied ex1-13 and CF-239 moved 1,200 feet (365.8 m) northeast to nest in territory III, having passed through two or three active territories on the way. CF-247, trapped in January 1967 in territory VIII, remained there to nest that year. By 1968, however, she had moved 500 feet (152.4 m) south into extralimital territory ex1-9. A new pair possessed VIII. CF-248, trapped in IX, nested there in 1967. In 1968, her former mate, CM-250, now with a noband female, moved north just beyond the boundary fence, to nest in extralimital territory ex1-16, leaving IX in possession of another winter-banded pair. After the first brood attempt in ex1-16, and the disappearance of the noband female, CF-248 reappeared and joined CM-250 again. Upon the completion of the second brood, they returned to IX to nest in the western portion. Before their nestlings fledged, CM-250 disappeared; the female of the pair in the eastern half of IX also disappeared. Unfortunately, there is no answer to the intriguing question of whether this was a simultaneous loss or a double desertion. CF-248 now paired with the remaining male in another nesting attempt.

In summary, three adult females, CF-86, 239, and 247, changed their territories, apparently because their mates were lost, and the new males attracted other females. The first two also found new mates, for, in their new territories, the former pairs had disappeared. The third female replaced a noband, and joined the resident male of the preceding year. CF-174 merely expanded her range into an adjacent vacant territory. CF-248's absence, while a rival displaced her for the first brood, her reappearance to join her former mate after the rival disappeared, her subsequent nesting in a territory adjacent to her old one, and her later return to her old territory, all suggest that her behavior depended upon that of her 1967 mate, CM-250. In 1968 he moved into ex1-16, probably because he was forced to. There he attracted another female perhaps more aggressive than his previous one. CF-248 returned when the vacancy developed and nested in ex1-16. Later she followed her mate into the more familiar territory IX.

In the course of the 6 years, 297 (70 percent) Cactus Wren nestlings of an estimated total of 424 were banded in the 49-acre (19.6 ha) research area. With the discovery in 1965 that one of the 1964 nestlings had established an extralimital territory, ex1-5, close to the east boundary, the decision was made to band as many of the nestlings of adjacent extralimital territories as time would permit. Beginning in 1966, 92 (80.7 percent) of an estimated total of 114 nestlings were so banded. Out of this grand total of 389 banded nestlings, only 27 males and six females survived to breed in the vicinity. Five of the males and two of the females had fledged in extralimital territories.

Six of the 27 male nestlings, in the following year, replaced their missing fathers in the same territories. Seventeen established new adjacent territories, evidently because the old territories were still occupied by their parents. One of these, CM-28, was absent the first year, possibly nesting extralimitally. Three left their old territories, occupied by their parents, and replaced lost males in adjacent territories. One established an adjacent territory, leaving his old one in possession of his brother. The average dispersal distance of 21 nestlings from the nest in which they had been hatched to their first breeding nest in a new territory was 370 feet (112.8 m); the minimum was 179 feet (54.6 m), the maximum 733 feet (223.4 m). The six wrens that did not change territories nested an average of approximately 100 feet (30.5 m) away from their old nest; the minimum was 29 feet (8.8 m), the maximum 139 feet (42.4 m).

The dispersal of the six surviving nestling females is more involved and will be described in greater detail in an attempt to account for their behavior. Only one, CF-121, fledged in territory V in 1965, remained to breed in her home territory. Her parents had disappeared, and CM-57, who had occupied the adjacent territory XVII with a noband female in 1965, vacated it after the loss of his mate, took possession of territory V, and paired with CF-121. Their first nesting attempt failed. When territory IV became vacant later in the spring, they moved in for their second brood, and held the territory at least through the summer of 1968.

Four other female Cactus Wrens left because their home territories remained in possession of their parents. CF-66, fledged in territory VII in 1964, found a mate, CM-54, a 1964 fledgling from territory I, and settled down for 2 years in a new territory, XVI (Fig. 17.2), adjacent to territory I and 1,100 feet (335.3 m) northwestward of her first home. Meanwhile, the parents of CM-54 continued to hold territory I. In 1967, CF-66 moved into the now vacant territory I, nesting there also in 1968. She had a new mate, CM-189, a nestling banded in territory I in 1966, who had taken possession of his home area when his parents disappeared. Her former mate, CM-54, and a female, CF-86, banded in territory XVI in January 1965, but not detected again until 1967, occupied extralimital territory ex1-11, north of territory I. The following year, 1968, CM-54 could not be found. CM-284, raised in 1967 in territory I by CM-189 and CF-66, now assumed control of ex1-11, aided by a noband female.

CF-230, fledged in territory VII in 1966, paired the following year with a noband male in extralimital territory ex1-8, about 1,400 feet (426.7 m) southwest of territory VII. In 1968, she could not be found. CF-234, fledged in territory IV in 1966, joined a noband male to occupy the adjacent territory V in 1967, vacated by her parents in 1966 when they moved into IV. Her first and only breeding nest was constructed 232 feet (70.7 m) from the nest in which she had been hatched. CF-238, fledged in 1966 in extralimital territory ex1-1 at the southeast corner, moved 185 feet (56.4 m) northwest to join CM-229 in establishing new territory XXI in 1967 (Figs. 6.8, 6.9). Her mate, CM-229, was a 1966 fledgling from territory XIII. Their first nesting attempt failed, but their 1968 efforts were successful. CF-265, fledged in 1967 in extralimital territory ex1-11, nested in 1968 in a new extralimital territory, ex1-12, 760 feet (231.6 m) to the east and 300 feet (91.4 m) north of the fence boundary. (A new male had taken possession of her home territory.) Her mate, CM-311, a winter-banded wren trapped in territory IX, may have been fledged in IX. The nestlings of the second brood in that territory were not banded.

The foregoing accounts suggest strongly that dispersal of the immature male wren is affected by parental antagonism in the circumscribed and defended territory. Presumably this antagonism begins sometime after the autumnal molt, or perhaps not until late winter, when territorial rivalries increase, for the juvenile wrens continue to occupy their roosting nests in the vicinity into the winter. The immature male's attachment to his home is great. He keeps possession of it if his parents are lost, but if they survive, then his first adult song probably brings on an immediate challenging response from his father, a threat which soon cannot be ignored. It appears evident that two males cannot occupy the same area during the breeding season. The immature males, less experienced, and probably less aggressive, left, but they retreated only to the point where the threat ceased, at the territorial boundary. Here they established new territories, or took possession of vacant ones. All the banded male nestlings that survived nested either in their old territories, or in adjacent ones.

The data relating to immature female wrens appear too meager and incomplete to warrant any calculation of percentages or averages. They do suggest that females disperse farther than males, and that males without mates are the attractive forces which encourage emigration, once the increasing hostility threatens the daughter with eviction. There is no evidence yet that a first year female wren can displace her mother.

Nineteen Cactus Wren pairs in the Saguaro Monument research area attempted only one nesting in a single season. Any change of mates here would be difficult to detect, unless the routine of laying, incubation, and feeding of nestlings was severely interrupted or delayed. However, 16 of these attempts involved banded males and five of them banded females. No changes or interruptions occurred. A different situation arose in the case of the 59 multiple broods, for replacements of losses or desertions could take place in the interval between nestings, and, if the wrens were not banded, the change would easily escape notice. Again the probability of not detecting a midseason substitution among the males was very slight, for 53 of the 59 were banded. In 1965, a banded male wren disappeared at about the time his first brood fledged. A noband replaced him for the second brood. A less assuring prediction applies to the females. Only 24 of the multi-brooded ones were banded. In 1968 a banded female deserted (?) when her nest was robbed of its nestlings. A noband female appeared and raised the succeeding two broods. These are the only replacements observed, except for the previously mentioned peculiar shifts in territory IX in 1968, under female dispersal.

In each of the 6 years, the research area probably contained the maximum available number of pairs of Cactus Wrens. Evidently the sex ratio was very close to 1:1, for unmated wrens were rarely detected. Two males, CM-84 and CM-245, established and maintained territories, but were unable to attract mates. The former, trapped in quadrat E2 in November 1964, built two nests in quadrat D2 in the spring of 1965 (territory XX, Fig. 6.6), meanwhile singing unsuccessfully from nearby trees in the constricted territory. Both of his nests were destroyed in the early autumn; he began another in October. By December, however, he had wandered as far southeast as G5, where he was trapped again. He then disappeared. CM-245, trapped outside the south boundary in December 1966, established new extralimital territory ex1-10, and worked on four nests in the course of the spring of 1967. No mate appeared. By early 1968, he had abandoned this small territory and moved north, taking over territory VIII, whose 1967 occupants had disappeared. He paired with CF-86, who had come from the far north end of the research area. Both of these unmated males had remained in their respective territories for the entire breeding season, waiting in vain for mates, instead of searching elsewhere.

It appears unlikely that any female Cactus Wren could hold a territory of her own. She would face the impossible task of remaining inconspicuous and silent. If she proclaimed her approach by song, an adjacent male would immediately investigate the challenge. Then would follow persistent harassment by his mate, as we observed on Kleindale Road, until the undesired female left. Several isolated, peripheral roosting nests in the contiguous creosote bush habitat appeared in use, and may have been occupied by female Cactus Wrens that had not succeeded in obtaining mates. However, it is improbable that they occupied these nests for very long in the course of the breeding season. Since the location of the male is fixed, a female must disperse if she is to mate. Enough evidence has been presented to show that females travel considerable distances to obtain mates.

18. Interspecific Relationships: The Curve-billed Thrasher

The Cactus Wren's distribution in southern Arizona coincides in large part with that of the permanent resident Curve-billed Thrasher. At least one pair of these thrashers was always present in our study area on Kleindale Road. They occupied a territory approximately the same size as that of the Cactus Wren pair. The spines of the cholla cacti presented no obstacles to them; they roosted exclusively in these cacti, and always chose them for their nest sites. As with the wrens, if a second pair crowded in, the territory of the first pair shrank to accommodate the newcomer at the margin.

Conflicts between the two species occurred frequently, and in many instances these conflicts seemed to be purposeful. At other times the behavior of the thrashers appeared erratic, confusing, and unpredictable. While feeding on the ground, the Cactus Wrens always gave way at the approach of the larger thrasher. No wren ever engaged in actual physical combat to retain its food supply. A short threatening run of a thrasher toward a wren sufficed; we never observed a thrasher attempt to drive a wren out of the thrasher's territory.

Fewer conflicts were noted at roosting time, for the Cactus Wrens retired early into their nests, usually at least 10 minutes before the thrashers retired. The latter often roosted regularly in the same chollas, sometimes only a foot or two from the wren's nest. No objections were raised by the wrens. Occasionally one appeared at the entrance of its roosting nest to investigate some disturbing noise made by a thrasher as it settled down for the night on a nearby twig. Satisfied that no threat was intended, the wren crept back into its nest. Although the thrashers must have been aware that the roosting nests were occupied, they attempted no interference at these times.

In the course of the breeding season, both species vigorously defended their own breeding nests. Thrashers chased wrens and wrens chased thrashers whenever the need arose; each was successful in defense in all conflicts we observed. Neither species damaged or destroyed the other's breeding nest at this time. When fledglings appeared, the clashes became more numerous. Cactus Wrens did not hesitate to attack fledgling thrashers which strayed into their vicinity. Not only did they chase them; they pecked them frequently, usually on the head; and the ensuing squeals brought the adults rapidly into a furious combat which raged from cholla to cholla until the thrashers retreated to a proper distance. Fledgling Cactus Wrens seemed to be more active and wary of danger; they seldom precipitated a conflict with the thrashers. Each species successfully defended its own young in its territory.

By far the most obvious evidence of conflict was the persistent destruction of Cactus Wren roosting nests by Curve-billed Thrashers. We first observed this puzzling, erratic behavior in the winter of 1932 (Anderson, 1934). Before this, we had attributed the occasional damaged nest to depredations of small boys who roamed the neighborhood. Since then, however, we have caught the thrashers in the act so frequently that there is no longer any doubt they were responsible for almost all of the destruction. Edwards (1919: 66) reported that "on the Mohave and Colorado deserts, and particularly the Mexican deserts, the large desert wood rats and ground squirrels cause the destruction of many nests." The type of nest, whether breeding or roosting nest, was not mentioned; neither was any actual proof offered that mammals were involved. Wood Rats were absent in our study area, but Round-tailed Ground Squirrels, Antelope Ground Squirrels, and Merriam Kangaroo Rats (*Dipodomys merriami*) inhabited the tract at the beginning. Later they vanished, perhaps because of the numerous cats in the neighborhood. We see no reason why any of these species should destroy a nest which presents no difficulty in entering. Furthermore, roosting nests do not contain eggs. Nevertheless, the destruction in our vicinity continued year after year, even after the rodents disappeared.

The pattern of nest destruction was nearly always the same; only the extent of the damage varied. Usually the thrasher began at the entrance of the roosting nest. It tore out, bit by bit, as much of the vestibule grasses as its bill would hold; then it dropped the pieces to its right and left. The whole operation suggested the alternate side strokes with which the thrasher digs into the ground in searching for food, but the movements were slower and more deliberate. Now and then it had to brace its feet to pull. Next went the roof of the nest, followed by the miscellaneous trash of the nest cavity, such as cotton, feathers, lint, and scraps of paper. The nests were often of flimsy construction; they could be torn apart with little effort. In every case the appearance of the destroyed nest was typical;

only a ragged cup remained. Sometimes only part of a nest was torn out, indicating, perhaps, that the thrasher had been interrupted or disturbed at its work.

These acts of destruction were not particular idiosyncrasies of individual thrashers. From 1940 to 1961, we trapped and color-banded most of the resident thrashers, a total of 30 individuals, and found that all of them acted alike. Both males and females destroyed the roosting nests of Cactus Wrens that had built in the territory of the thrashers.

We recorded approximately 200 instances of total or partial destruction of Cactus Wren nests in our study area from 1932 to 1961. There are many gaps in this record; the actual total of nests damaged must have been considerably greater. Damaged roosting nests numbered 160; damaged secondary nests numbered 11. Thrashers did not attack breeding nests while they were in use. After the young had fledged, however, they tore apart 18 of these nests. In addition, eight nests, abandoned following unsuccessful nestings, suffered similar damage. Nests were not always completely destroyed in the initial attack; 25 nests were damaged twice, seven nests three times, four nests four times, and one nest seven times, before the destruction ceased.

We found nest destruction to be erratic. Sometimes a roosting nest would remain undisturbed for weeks, or even months; another nest would be ripped apart as soon as it was completed; others were damaged while under construction. Particularly vulnerable were unoccupied nests in chollas in which thrashers roosted. If a thrasher roosted beneath such a nest, it would tear it open at the bottom. The destruction decreased to a minimum in the breeding season, from February to May. It increased in June, and then continued at a rate two to four times as high throughout the summer and autumn until January. Although there appeared to be a definite increase in territorial assertion in the fall of the year, as exemplified in this orgy of nest destruction, the wrens were not otherwise molested.

Despite the evident importance of roosting nests, Cactus Wrens never defended them against attacks from thrashers. We saw disputes occasionally among wrens themselves for possession of a nest at roosting time, usually when the fledglings approached independence, and now and then we noted that they even went so far as to eject one of their own kind that had usurped a nest. It seems incredible that the wrens could be unaware of the nest destruction. They foraged regularly in the vicinity when it occurred, and they must have witnessed an act so conspicuous; yet, at no time did we see one oppose the thrasher, or utter a *tek* or *buzz* note in protest. Furthermore, after the departure of the thrasher, the wren ignored the damaged nest.

Not until evening did the disturbing picture of destruction confront the wrens; then it apparently took them by surprise. When one landed on the doorstep, it stood upright in actual bewilderment. Then it leaned forward as though to enter, but it could find no opening, for the nest was now a ragged, shallow saucer. Sometimes it gave up the attempt, and flew away in search of a vacant nest. If all were occupied, it returned and tried to enter. Finally, after considerable moving about and *buzzing* in frustration, it settled down into the untidy floor of feathers and hid its head in the fluffy mass, its back exposed to the sky. Fledglings and immature wrens continued to roost in such nests until the structure flattened, fell apart, or otherwise became uninhabitable. Adults might occupy the nest for as long as a week; then they constructed new nests for roosting. The owner seldom endeavored to repair a damaged nest. If a nest was under construction at the time it was damaged, it was sometimes repaired, provided the damage was not extensive. Frequently the first attack was only the beginning, for the thrasher returned in a day or two to finish the job. Eventually the nest was abandoned. Later, a new tenant often added material and made an effort to install a new roof.

Curve-billed Thrashers made no use of nests they destroyed. Rarely, an adult or a fledgling could be found roosting on the floor of an abandoned wren's nest, but as a rule, they chose the horizontal branches of the chollas, close to the main trunk. In the winter months, thrashers often carried small twigs to their roosting chollas to be fashioned into a loose platform on which they sometimes roosted. Later, this platform might become their breeding nest.

The distance between nests of the two species in the Kleindale Road area ranged from an incredible 6 inches (15 cm) in 1970, and 3 feet (.9 m) in 1960, to a comfortable 480 feet (144 m) in the year of maximum separation. Nests with first clutches averaged 177 feet (53.1 m) apart; nests with second clutches averaged 118 feet (35.4 m) apart. The only two nests with third clutches were 210 and 120 feet (63 and 36 m), respectively, from nests of Cactus Wrens with third clutches. Nest sites chosen, of course, depended upon the availability of the cholla cacti, and these were very irregularly spaced.

The 3-foot separation of nests occurred in 1960. Unfortunately these nests were in cholla 67, at the edge of Flanwill Street, completely hidden from our view by shrubbery and buildings on the intervening lot. We obtained no details of any conflicts there. The first Cactus Wren egg was laid about 13 April; that of the Curve-billed Thrasher was laid about 30 April. Both species fledged successfully, the wrens about 21 May, and the thrashers about 29 May. Apparently the Cactus Wrens recognized the dangerous situation, for they made no attempt to lead their fledglings to roost in the old breeding nest. Instead they chose another nest at a safe distance in our front yard. A week later their abandoned breeding nest had been torn slightly; eventually it was destroyed.

By the middle of February 1970, the wrens had completed nest 5T in lot 7; the male occupied a roosting nest in cholla 77 near the north fence. Meanwhile, a pair of thrashers outlined a nest in a cholla near the northeast corner of our house. Soon they started another nest less than 2 feet (.6 m) away. When this second nest appeared completed, they abandoned it, and began another nest in cholla 5, below and south of the

Fig. 18.1.　Cactus Wren and Curve-billed Thrasher nests, April 1970, in cane cholla in Lot 7 on Kleindale Road, Tucson, Arizona.

Cactus Wrens' nest, 5T. Its rim of twigs finally extended upward to within 6 inches (15 cm) of the rounded bowl of the wrens' nest (Fig. 18.1). The entrance of the latter faced north, away from the thrashers' nest.

The wrens had four eggs in nest 5T on 6 March. While the wren incubated, unmolested, above, the thrashers continued building below. By 10 March, the lining was in place, and on 20 March the thrasher laid her first egg; she laid two more. On 22 March, the wrens were carrying food rapidly and repeatedly to their four nestlings, seldom pausing even a moment on the doorstep before entering. Below, the thrashers incubated, alternately, neither species crossing the line of flight of the other, and both apparently indifferent to or unconcerned about the other. On 3 April, one of the thrasher eggs hatched; another hatched the following day. Brooding and feeding of the two nestlings continued. On 8 April, the first Cactus Wren nestling fledged. Our curiosity led us to cause the accidental fledging of two more the same day. In the forenoon there was a brief encounter at the east fence with a thrasher, and a short chase ensued. About 1815, the Cactus Wrens and their fledglings drew nearer to nest 5T. A thrasher suddenly landed on an upright twig above the nest. At once, an adult wren attacked it so explosively that the startled thrasher rose vertically, unbalanced, and fluttering in apparent confusion. As they dropped to the ground, they separated. A few minutes later, the wrens began leading their fledglings into nest 5T. All three entered for the night. A thrasher on cholla 25, 20 feet (6 m) away, watched the entire retirement ritual; the other thrasher visited its nest once with food, and when the third Cactus Wren fledgling retired, the thrasher came again to feed its nestlings below. There was no conflict at either visit. For a short while, both species carried food to their nests. Once, they arrived and departed simultaneously. On 9 April, nest 5T was vacant, but the fourth nestling could not be found. In the evening, at least one fledgling retired in nest 5T, and another in the male's roosting nest in cholla 77. The next evening, with less coaxing, all three fledglings retired in the male's nest. The male, apparently, was forced to roost in the secondary nest in cholla 67 on Flanwill Street. The female flew southwest, leaving nest 5T vacant. On 20 April, the two thrasher nestlings fledged. Cactus Wren nest 5T, still vacant, remained intact.

In both of the above close nestings, the wrens had the advantage of prior ownership in the cholla location. They were already firmly established when the thrashers began construction of their "basement" apartment. Had the two species started nest building simultaneously, in such intimate proximity, the thrashers would probably have dominated the location, forcing the wrens to leave.

Dates of first eggs laid in our vicinity varied considerably. Now and then the two species laid at almost the same time; in some years, the wrens were first; in others, the thrashers were first. The young fledged sufficiently close together to be in competition for food. Generally, in a given season, Cactus Wrens attempted to raise a greater number of broods. In addition, their clutches averaged larger. In 12 of the 22 years from 1939 to 1960 we have reasonably complete nesting data on both species. The Cactus Wrens fledged 82 young from 27 successful broods, with an average of 6.83 per year; the thrashers fledged 51 young from 25 successful broods, with an average of 4.75 per year. In spite of their lower productivity, the thrashers maintained their relative numbers in the study area. It is difficult to draw any safe conclusion here, for we really know very little of the dispersal into adjacent territories or of the immigration of outside members of the two species to our area. We suspect that it was the outsiders who filled the gaps occasioned by local losses of birds. The depredations of the thrashers seemed to have little effect upon the size of the wren population; it remained stable also.

Both Huey (1942: 368) and Hensley (1954: 200) have mentioned competition for nest sites of these species in Organ Pipe Cactus National Monument in southwestern Arizona. The former reported that "their choice habitat amongst the cholla cactus was occupied commonly by Palmer Thrashers and the competition appeared to be too much for the wrens." However, he gave no figures on relative abundance or observations on definite conflicts. Hensley likewise stressed the competition, saying that "the denser more luxuriant chollas were usually taken by the thrashers which needed considerable protection for their nests." Our studies have not yet proven that thrashers need denser chollas or more protection for their nests than other birds in the area. On our lot, Curve-billed Thrashers nested frequently in the cane cholla, a cholla of open structure. The mere fact that a thrasher chooses a particular cholla for its nest does not seem to be sufficient proof that a conflict occurred which resulted in a Cactus Wren being forced to accept an "inferior" location. In an undisturbed habitat, many dense chollas are not occupied by either wrens or thrashers. The thrashers avoided some competition by using the same nest two or more times for breeding purposes. Hensley's figures (1954: 195) on population density actually indicate that the Cactus Wrens were more abundant than the thrashers in each of the three areas he studied.

We found many torn or destroyed nests on the Santa Rita Experimental Range that were typical of Curve-billed Thrashers' work. In their nesting, the thrashers did not exhibit the same drastic changes that occurred in the wren population. The thrashers built four first brood nests in 1953, all of them in wren territories (we have no data on second broods for this year). One of the nesting attempts failed. In 1954, we located three first brood and three second brood nests. Two of the latter failed. Three of the nests were in areas not occupied by wrens. In 1955, they built seven first brood nests, four of which were abandoned, two before eggs were laid, and two afterward. Later they attempted to raise two second broods; one of these failed. Only two nests in that year were in Cactus Wren territories. In 1956, five first brood nests contained eggs; two nests were in wren territories. Curve-billed Thrashers constructed all of their nests in cholla cacti, usually within the framework of the spiny joints. They appeared to be well protected from the larger mammalian and avian predators, but not from snakes or Roadrunners.

The Saguaro National Monument research plot offered an excellent opportunity to devote particular attention to the assumed competitive activities of Curve-billed Thrashers. Our Kleindale Road observations had been far from conclusive. The results of banding adults and nestlings, and observations of nests in the larger Monument area, are reported in detail here.

Curve-billed Thrashers established fewer territories each year than Cactus Wrens, and, although the annual variations were far less spectacular (Fig. 6.3), they did exhibit a weak tendency to parallel those of the wrens. From 1963 to 1968, the number of territories were 5, 6, 8, 7, 7, and 7, respectively, with an average of 6.7. Several adjacent extra-limital territories are included in five of the years, because the thrashers in these territories extended their activities into the Monument research

Table 18.1. Loss of winter-banded Curve-billed Thrashers in
Saguaro National Monument.

Banding period, winter	Number banded	Number nested	Number lost	Percentage lost
1962-1963	6	6	0	0.0
1963-1964	3	3	0	0.0
1964-1965	3	3	0	0.0
1965-1966	6	5	1	16.7
1966-1967	2	2	0	0.0
1967-1968	9	4	5	35.6
Total	29	23	6	20.7
Average	4.8	3.8	1.0	

area, but nested a short distance outside. If, however, these nests are excluded, the number of territories reduces to 4, 6, 6, 4, 4, and 5, respectively, with an average of 4.8.

The territorial designations A through J do not correspond to, nor are they equivalent to those of Cactus Wrens. In most cases they included and overlapped several Cactus Wren territories. For example, territory A spread irregularly over territories I, II, IV, V, XVII, XX, and XXIII of the Cactus Wrens.

The 6-year average of 3.8 banded, breeding Curve-billed Thrashers (Table 18.1) was disappointing, barely exceeding half the average of 6.7 territories per year. Despite the good, overall male to female trapping ratio — far better than that of the Cactus Wrens — it turned out that four out of five territories had noband females in 1963, four out of six in 1964, four out of eight in 1965, two out of seven in 1966, and three out of seven in 1967 and 1968.

The locations of the five territories established in 1963 remained relatively stable in the course of the following 5 years (Figs. 18.2 to 18.7). When territories F, G, and H intruded in 1965, they forced a slight southward displacement of territory A; and, in 1968, the pair in C moved southwestward, giving up their old extralimital nesting area to a new pair. With the exceptions of territories C and E, whose males survived the entire period of 6 years, the territorial allocations may be somewhat arbitrary.

The average available space for each territory of the Curve-billed Thrashers was about 10 acres (4 ha), but it is doubtful if each pair utilized or defended more than half of this area regularly. Attacks were vigorous against thrashers who intruded on the vicinity of their nests. In all, 30 chases were recorded, some of them extending as far as 400 feet (121.9 m). Other chases, ending in furious combat, occurred on 22 days. Many

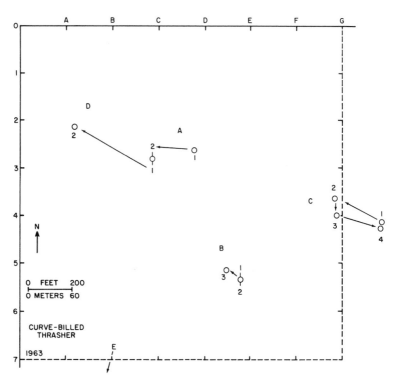

Fig. 18.2. 1963 Curve-billed Thrasher breeding nests in Saguaro National Monument. The nest in territory E was extralimital, but the thrashers ranged north into Quadrat B6. The pair in A took over the vacant first brood nest of D for their second clutch, as B moved westward.

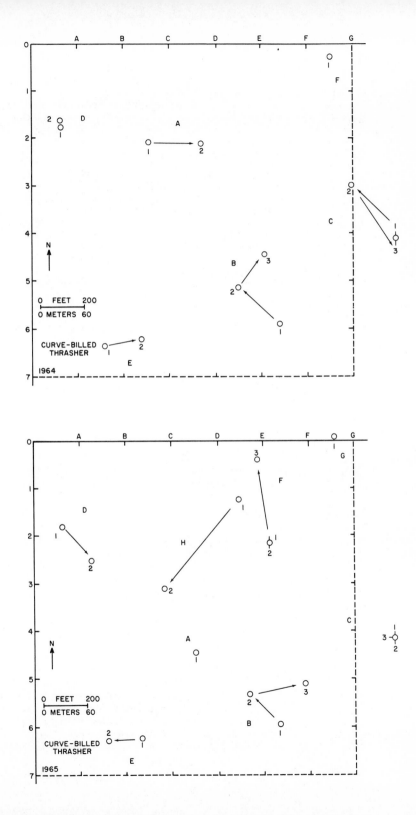

Fig. 18.6.   1967 Curve-billed Thrasher breeding nests
in Saguaro National Monument. No change in
territories E, F, and G. B has moved westward;
C has split into two uncertain divisions —
owners not identified.

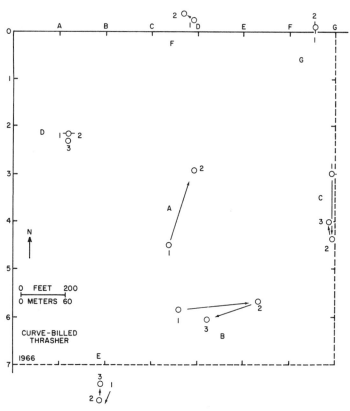

Fig. 18.5.   1966 Curve-billed Thrasher breeding nests
in Saguaro National Monument. Territories E, F, and G
are astride the boundary; C is again inside.

Fig. 18.7.   1968 Curve-billed Thrasher breeding nests
in Saguaro National Monument. Territory C is now
displaced into the southeast corner, making room for a
new Territory J. B and A have joined, after the 1967 male
in A disappeared; G was abandoned before egg-laying.

[178]

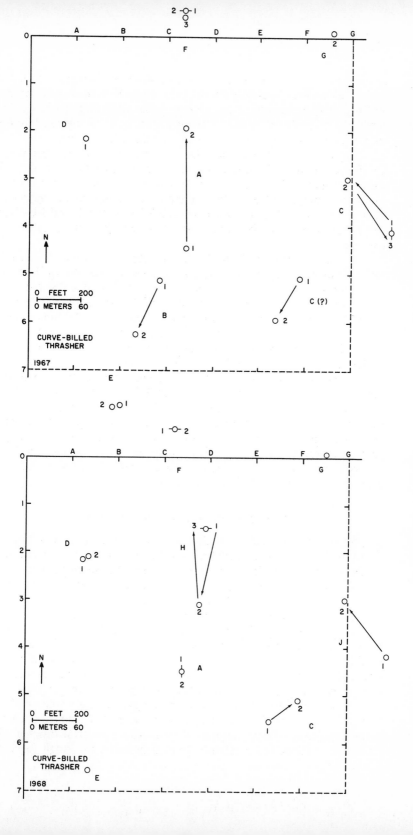

more probably took place. Four were observed in January, six in February, ten in March, eight each in April, May, and June, and one each in the remaining months of the year, with the exception of three in September. Curiously, a combat usually ended with the defeated retiring only a short distance, when the victor refrained from pursuit.

Physical combat resembled that of domestic roosters, but was of shorter duration. Both thrashers rose vertically from the ground, sometimes as high as 6 feet (1.8 m), while they pecked and fluttered excitedly in close combat. Falling to the ground, they continued the battle in a confused tangle, apparently grasping each other with bills and feet. Upon separation, the vanquished retreated, often with its bill open from the exertion.

Conflicts occurred chiefly in the vicinity of nests, not at the territorial boundaries, in contrast to those between Cactus Wrens. In March 1966, the two banded thrashers in territory A chased a noband round and round their completed nest for at least 15 minutes, finally driving it away. The chase included a fight on the ground. Usually males chased males, but at times, when the pursuit involved three thrashers, it was impossible to determine whether one of them merely followed in the excitement. On 12 June 1967, a thrasher, presumably a female, attempted to force its way into a nest in territory E, already occupied by an incubating thrasher. A chase ensued in the cholla, followed by the intruder's departure, the owner of the nest in rapid pursuit. This suggests an unbalanced sex ratio, a condition more noticeable on Kleindale Road in the spring of 1958. Such attempts there were frequent, persistent, and actually successful, resulting in two instances of polygyny. When a conflict occurred between two females over a nest containing eggs, the male remained at a distance, apparently only an observer!

Nesting of the Curve-billed Thrasher

In direct competition with Cactus Wrens, the Curve-billed Thrashers chose the jumping cholla for 54 of their 61 breeding nests in the Saguaro Monument research plot, and for 27 of their 32 nests in the contiguous extralimital area. They chose the staghorn cholla for six nests inside the area and five nests outside. Nest heights varied from 2.5 feet (.76 m) to 6.5 feet (1.98 m) in the jumping cholla, and from 6 to 7 feet (1.8 to 2.1 m) in the staghorn cholla. In the jumping chollas, they placed some nests close to the trunk and others in the outer branches. All had some shade; none were in complete shade. The size and shape of the chollas obviously influenced the choice of nest location. Nests in staghorn chollas were usually in the center of the plant. A saguaro stump, 10 feet (3 m) high, held one nest built upon the pulpy interior, entirely without shade. Several other stumps in the area, 7 to 9 inches (17.5 to 22.5 cm) in diameter, contained thin platforms of twigs placed upon the disintegrating pulp. These abandoned, unfinished nests appeared too small to be the

Table 18.2. Date of first Curve-billed Thrasher egg,
Saguaro National Monument.

Year	Number of nests	Average Date 1st egg	Range	Standard deviation
1963	3	27-2	22-2—2-3	5.9
1964	6	26-4	8-4—25-5	37.7
1965	7	16-4	14-3—26-5	67.1
1966	6	16-3	14-3—18-3	3.5
1967	8	27-4	24-3—15-5	46.4
1968	10	24-2	11-2—22-3	36.3

work of Roadrunners who sometimes constructed nests in roomier saguaro stumps.

Most of the thrasher nests were new; a few were nests of the previous season which received new lining. Five nests were used again for second clutches; one served for three clutches. They shared this economy of nest use with the Cactus Wrens. Rarely they placed a nest upon a flattened wren's nest. Wrens balanced the score by constructing roosting nests upon old, dilapidated thrasher nests. At no time, however, were active breeding nests of the two species placed in the same cholla. On 9 March 1965, two thrashers worked on a nest in a cholla in quadrat A2. By the 17th it had lining, and 2 days later it was occupied. Meanwhile, a Cactus Wren laid four eggs, the first on 11 March, in a nest only 3 feet (.9 m) from the thrasher nest. The thrashers deserted their nest before egg-laying; the wrens fledged three young. The average distance of 28 thrasher nests from the nearest Cactus Wren breeding nest was 210 feet (64 m); the range was 72 to 410 feet (21.9 to 125 m). The lowest average distance apart, 158 feet (48.2 m), occurred in 1965, when the wrens established their maximum of 20 territories.

Table 18.2 gives the average dates of the first eggs laid. In four of the years, the Curve-billed Thrashers laid earlier than the Cactus Wrens (Table 8.3), but two late nestings in 1964 and four in 1965 are obviously delayed ones; omitting these, all Curve-billed Thrasher dates appear earlier than those of the wrens. Difficulty in finding mates and choosing nest sites may account for the wide variations in dates of laying.

The same temperature-precipitation conditions that apparently influenced the start of laying of the wrens may also have triggered that of the thrashers from 1964 to 1967, for the difference in time response did not exceed 8 days. In 1963, however, a steep rise in daily mean temperature occurred in the first week in February, followed by a sharp drop during three rainy days. Then, after a smaller, but uniform, rise in temperature, the thrashers began laying. The wrens apparently required a longer period of higher mean temperatures. The 1968 layings are more perplexing, for they actually began in a dip in the mean temperature. Two fragmentary records for the year 1962 are available. Estimated dates of

**Table 18.3. Number of Curve-billed Thrasher clutches laid,
Saguaro National Monument.**
Figures in parentheses indicate incomplete clutches.

Territory	1963	1964	1965	1966	1967	1968
A	2	2(1)	1	2	2	2
B	3	3	2	3	2	
C	4	3	3	3(1)	5(2)	2
D	2	2		3(1)	1	2
E	1	2	1	3	2	1
F		1	3(1)	2	3	2
G			1	2	2	
H			1			3
J						2
Total	12	13	12	18	17	14
Average	2.4	2.2	1.7	2.6	2.4	2.0

first eggs laid are 1 February in territory A and 3 February in territory E, the earliest in the 7 years. These dates are approximately 2 to 3 weeks ahead of those of the Cactus Wrens that year. For comparison, in the single territory on the Kleindale Road lot, from 1963 to 1968, the dates of first egg laid were 10 February, 16 March, 2 March, 13 February, 13 February, and 14 February, respectively.

The average number of clutches per year was lowest in 1965 (Table 18.3), the year the Cactus Wren population reached its maximum of 20 territories. Subtraction of incomplete clutches — those deserted or robbed after only one egg had been laid — reveals in the summary a remarkably close inverse relationship with the number of Cactus Wren territories. The average clutches are 2.4, 2.0, 1.6, 2.3, 2.0, and 2.0; the number of Cactus Wren territories totalled 9, 11, 20, 9, 16, and 16, respectively. This relationship is also apparent, though less striking, in the average eggs per set (Table 18.4). Again omitting the incomplete sets, the averages

**Table 18.4. Size of Curve-billed Thrasher clutches laid in
Saguaro National Monument.**
Figures in parentheses are incomplete clutches; they are included in totals.

Year	1 egg	2 eggs	3 eggs	4 eggs	Totals Nests	Totals Eggs	Av. per set
1963			11	1	12	37	3.1
1964	1(1)	1	11		13	36	2.8
1965	1(1)	3	8		12	31	2.6
1966	3(2)	3	9	3	18	48	2.7
1967	3(3)	7	7		17	38	2.2
1968		4	9	1	14	39	2.8
Total	8	18	55	5	86	229	2.7

are 3.1, 2.9, 2.7, 2.9, 2.5, and 2.8 eggs per set, the last figure being a departure from the inverse tendency. The increase from 2.5 eggs per set in 1967 to 2.8 in 1968 (12 percent) is considerably lower than the 3.0 to 3.6 (20 percent) increase in the wren production. In both species, the laying of larger clutches contributed to these figures. (In the single territory on the suburban Kleindale Road lot, 54 nests from 1939 to 1969 averaged 3.0 eggs per set.)

Because of the many failures, nesting season length has been calculated in some of the years from the date of the first egg laid to the last egg laid, instead of to the fledging of the last young. It was 136 days in 1963, 110 in 1964, 108 in 1965, 131 in 1966, 75 in 1967, and 179 in 1968, the last identical with the Cactus Wren season that year. The average length of the thrasher nesting season was 4.1 months, about 5 percent longer than that of the Cactus Wrens. If the late nests had been successful, the season would have been considerably longer.

Curve-billed Thrasher nesting success

The 20.9 percent nesting success of the Curve-billed Thrasher for the 6-year period seems incredibly low (Table 18.5). It is a minimum estimate; the fate of 28 of the 86 clutches is in doubt, for their banded nestlings could not be located shortly after they should have fledged. There were only vacant, undisturbed nests. If the nestlings in the 28 doubtful nests fledged successfully, the success rises to 43.2 percent, a highly

Table 18.5. Curve-billed Thrasher nesting success, Saguaro National Monument.

1963	1964	1965	1966	1967	1968	Total
			Clutches			
12	13	12	18	17	14	86
			Successful clutches			
4	2	3	5	2	5	21
			Percentage successful clutches			
33.3	15.4	25.0	27.8	11.8	35.7	24.4
			Eggs laid			
37	36(3)	31(3)	48	38	39	229
			Eggs hatched			
24(3)	19(3)	16	30	15	21(2)	125
			Percentage eggs hatched			
64.9	52.8	51.6	62.5	39.5	53.8	54.6
			Fledged			
9	5	5	16	3	10	48
			Percentage eggs fledged			
24.3	13.9	16.1	33.3	7.9	25.6	20.9
			Percentage hatch fledged			
37.5	26.3	31.3	53.3	20.0	47.6	38.4

improbable value, considering the complete disappearance of all the banded nestlings. (A tabulation of 54 nests from 1939 to 1969 in the single territory on suburban Kleindale Road indicates that 58.1 percent of the eggs fledged young.)

In the Saguaro Monument area, only 48 nestlings fledged out of 95 banded. Five extralimital nests, with 13 banded nestlings, are excluded from this total. In the 6 years, thrashers averaged 1.2 young fledged per pair. The annual loss of the adult population is not known. Assuming it averaged 40 percent (Nice, 1937: 177), the adult loss from 1963 to 1965 would be 4.0, 4.8, and 6.4, respectively, and 5.6 adults in each of the remaining 3 years. In 1965 and 1967, nestlings fledged did not exceed the losses of those years; in 1964, the gains and losses were about equal.

Thrashers deserted six nests before egg-laying began, and 16 nests containing eggs, some before the clutches were complete. Predators robbed 18 nests. One nestling died in its nest. On 9 to 10 March 1968, about 1.6 inches (41 mm) of cold rain killed the entire brood of three 5-day old nestlings in territory F, two 6-day old nestlings in territory C, and two 3-day old nestlings in territory A. This rainy period had no adverse effect on the better protected Cactus Wren nestlings.

Snakes were probably responsible for most of the egg and nestling losses. Thrashers uttered alarm notes when rattlesnakes approached, evidently recognizing them as enemies. The two diurnal ground squirrels would have difficulty in obtaining eggs, for both thrashers alternate in incubation. The same holds true in brooding. A passing coyote (*Canis latrans*) caused a thrasher to leave its nest to protest. Roadrunners are not above suspicion, in view of the great losses suffered by the recently fledged nestlings. These slow, inexperienced young birds would be easy prey on open ground. One can easily seize, by approaching slowly and quietly, fledglings dozing on cholla branches.

Two male Curve-billed Thrashers, TM-1 in territory C and TM-3 in territory E, nested in their respective territories for at least 6 years. Tenure of the other 27 banded adults exceeded 3 years in only one instance. Four winter-banded adults disappeared before the breeding season was under way. A fifth drifted northward beyond the boundary, remaining until April. A sixth, T-115, vanished after failing in its 1968 nesting attempt north of the area. Unfortunately, the record of eight thrashers is incomplete, since they were still present in 1968 when the study ended. Banded thrashers, absent for a season, are presumed dead; they could not be located again in the vicinity.

On the somewhat indefensible assumption that all the variables of both species were equal, the Curve-billed Thrashers suffered a loss of 20.7 percent of their winter-banded adults (Table 18.1), compared to 49.4 percent for the Cactus Wrens (Table 16.2). About half of the winter-banded wrens survived to breed in the following spring; four-fifths of the thrashers survived.

The disappearance of 47 of the 95 banded nestlings seemed cata-strophic. The fate of the remaining 48, which are presumed fledged, was little better. Twenty-eight could not be found, but since, at the last observa-tion, they appeared ready to leave their nests, they are counted as fledged. Nineteen others remained in their territories an average of only 15 days; the range was 3 to 25 days. One fledgling, T-43, banded 7 April 1965 in territory C, held on until 10 September, apparently not disturbed or threatened by its parents. None of the banded nestlings appeared in the winter trapping efforts, and not one nested in the research area or its immediate vicinity. Emigrants from outside filled vacancies in the breeding population. This surplus population, presumably in the neighborhood waiting for a vacancy, could not be detected. There was no interchange of banded individuals from or to contiguous territories. (In the course of the autumn of 1941, when thrashers in the Kleindale Road tract destroyed so many of the roosting nests of the Cactus Wrens that we feared the wrens would leave, we began trapping thrashers, some of which were banded, and releasing them 2 to 5 miles [3.2 to 8 km] distant. To our amazement, each time we returned home we discovered that the deportee had already been replaced by another. After five such futile attempts, we abandoned the experiment.) The possibility of greater dis-persal cannot be ruled out entirely. A nestling thrasher banded 17 February 1941 on Kleindale Road, present until 26 March 1941, was found dead 1 June 1941 in a residential district approximately 3 miles (4.8 km) south. Although such a dispersal in an urban area, fraught with many man-made hazards, is not comparable to that in the Monument, it does suggest that survival of fledglings for a considerable time, and movement to unexpected distances, can take place.

In Saguaro National Monument, severe conflicts between Curve-billed Thrashers and Cactus Wrens must have been rare. In the usual overlap of territories, there was apparently no attempt by either species to exclude, or even to threaten to evict, the other. The large numbers of cholla cacti in the Monument would seem to make conflicts unnecessary. Many apparently suitable chollas were never used; others served for nest sites year after year. A few of the more scattered, densely-branched jumping chollas appeared to be particularly attractive to the thrashers, which nested in them for several consecutive years. Other thrashers pre-ferred the relatively open-branched staghorn chollas. The extensive use of palo verde trees and saguaros for nest sites by the Cactus Wrens further reduced the possibilities of conflict. There is no evidence that they chose these locations because of thrasher interference in the cholla cacti. As on Kleindale Road and in the Santa Rita Experimental Range, destruction of Cactus Wren roosting nests by Curve-billed Thrashers was extensive (Fig. 18.8), but it did not appear to affect wrens' breeding activities. Nests with eggs or nestlings were spared. Most of the destruction began in the autumn while the molt was in progress, and continued until Decem-

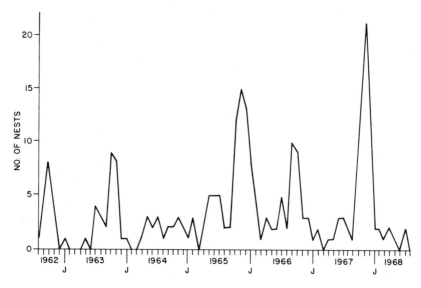

Fig. 18.8.   Cactus Wren nests in Saguaro National Monument damaged or destroyed by Curve-billed Thrashers from August, 1962, to August, 1968.

ber. The renewal of singing at this time may indicate an intensification of territorial activity. Thrashers did not destroy nests to make room for their own. They nested in other cholla cacti.

Curve-billed Thrashers were primarily ground feeders, never gleaning from branches of shrubs and trees like the wrens. Apparently their feet are not adapted for climbing. Thrashers did, however, have a wider range of food preferences than the Cactus Wrens, and also differences in their manner of acquiring food. Both species, in direct competition, foraged on the ground in autumn and winter, picking up food from the surface and from low vegetation within reach. In addition, at all times of the year, the thrashers dug frequently into the soil, making conical holes to depths of 2 inches (5 cm) or even more, to obtain buried arthropods. When walking about in the abundant litter, they exposed the ground by using their bills to pick up and toss aside the fallen cholla joints. Cactus Wrens pushed the smaller debris aside, or turned it over with their bills to facilitate their search for food. They did not dig. By the time nesting was well under way, deciduous perennial vegetation leafed out. The wrens spent less time now on the ground, and by midsummer they foraged frequently in the palo verde and mesquite trees, the cholla cacti and saguaros. The thrashers, whose weight averaged twice that of the wrens, could attack and capture food items of considerable size, thus reducing slightly the competition. Once, when a nestling was lifted up for banding, it disgorged a full-sized, slimy, cicada!

Thrashers ate gray-thorn berries in the spring, and the presence of the indigestible stones in vacated nests indicated that they had also fed them to their nestlings. They swallowed whole the fruits of Mexican crucillo and desert hackberry, both available in good supply from July to December, despite their use by other birds as well. The abundant, glutinous fruits of the leafless mistletoe, a plant widely parasitic on palo verde and mesquite trees in the Monument area, were apparently ignored. The red, ripe fruits of prickly pear gradually disappeared in the course of the autumn, but the thrashers shared this food with several other species of birds and rodents. Fruits of barrel cactus (*Echinocactus wislizenii*) provided seeds in the late fall and winter. The initial penetration of its outer skin and pulp may have been accomplished by the competing Gila Woodpeckers. Cactus Wrens were not observed eating any of the above berries or fruits.

While foraging on the ground, both species probably picked up various seeds. Judging from the vast numbers of winter ephemeral plants, which appear after good winter rains, the entire bird population must consume only an insignificant quantity of seeds. The most abundant plants were those whose fruits appear most unpalatable or difficult to swallow: for example, species of *Erodium, Pectocarya, Cryptantha,* and *Amsinckia*. Many of the summer plants, like the threeawn grasses (*Aristida* spp.) and the needle grama, appear equally unsatisfactory. Thrashers will eat seeds like millet in commercial bird foods, but they do not eat to excess. They have been seen leaving such food to a hungry, frantic group of House Sparrows, to resume digging nearby for insect food.

The conspicuous white flowers of the saguaros, usually opening in May, attracted both wrens and thrashers frequently. The birds probed their bills deep into the long corollas, apparently to obtain the nectar, present to a depth of 0.25 inch (.6 cm). They no doubt also captured any insects trapped in it. Both species probably assisted in cross pollination. Later, when the ripe fruits split open, both ate the moist, red pulp and the seeds. The flowers and fruits may be a supplementary source of water in the rainless months of May, June, and part of July. The extensive list of food items reported by Ambrose (1963: 24-30) indicates that the thrashers are almost omnivorous.

19. Interspecific Relationships: Other Birds, and Predation

Few conflicts could be detected between Cactus Wrens and the varying numbers of migrants arriving in the fall to remain for the winter. Brewer Sparrows, White-crowned Sparrows, and Lark Buntings, all seed eaters, made inroads on the available food supply, but apparently there was enough food for all. Some of these birds roosted in cholla cacti that held wrens' roosting nests. By retiring later than the wrens, they avoided detection and possible interference. Wintering Mockingbirds (*Mimus polyglottos*) and Phainopeplas (*Phainopepla nitens*) fed regularly, undisturbed, on the mistletoe berries in our large mesquite tree.

Transients from adjacent Rillito Creek streamside vegetation, such as Brown Towhees (*Pipilo fuscus*), Rock Wrens (*Salpinctes obsoletus*), and even Cardinals (*Richmondena cardinalis*), were occasionally chased, but not very far. A Western Kingbird (*Tyrannus verticalis*), that had built a nest on the wire bracket of an electric pole located near a wren's breeding nest, prevented the wren from using the pole as a singing station. On 24 February 1957, we saw a male wren singing from the tall television antenna in our neighbor's lot. A few feet away, a male Pyrrhuloxia alternated with his vigorous, whistled song. Singing Mockingbirds and Cardinals were not molested. Cardinals that brought their fledglings into our vicinity were once observed chasing several immature wrens from a mesquite tree.

The species that nested commonly in the same area as the Cactus Wrens were: Mourning Dove, Inca Dove (*Scardafella inca*), Curve-billed Thrasher (discussed in Chapter 18), House Sparrow, and House Finch. The Inca Dove and House Sparrow, both dooryard birds, are not found in the usual habitat of Cactus Wrens. More accurately, one could say the Cactus Wrens had remained here until their marginal area had become semi-domesticated. The desert, except for small patches like those in our back lot, gradually disappeared as the human population moved northward to the bank of Rillito Creek.

Mourning Doves, which had seldom nested in our area before, have become more numerous in the past 15 years. Freedom from disturbance, and our constantly filled pool of fresh water, no doubt contributed to their desire to stay. They nested commonly in cholla cacti, often directly upon an old Cactus Wren's nest. There never was a serious question of ownership. At the approach of a wren, the incubating Mourning Dove raised up threateningly. That was enough; the wren backed up and left. In the spring of 1961, when the Cactus Wrens left our lot to breed in adjacent lots, three pairs of Mourning Doves moved in. In all, they made 14 nesting attempts that year, lasting from February to September. Twelve of the nests were in cholla cacti. Some of the nests were used over and over again, thus reducing the competition for new sites. About half the nesting attempts were successful. So far as we could observe, none of the losses of eggs or nestlings, or desertions, could be attributed to conflicts with Cactus Wrens.

Although in direct competition for nesting sites in cholla cacti, Inca Doves appeared to have little difficulty in maintaining their hold in the vicinity. Nesting success has been summarized through 1947 (Anderson and Anderson, 1948). We observed few conflicts with Cactus Wrens. Two doves, that began building a nest in a small cholla, abandoned the attempt when they were chased several times by a male wren. We once saw a wren drive away a dove that had landed on a roosting nest. At other times they fared much better. Twice the Inca Doves succeeded in fledging their broods from nests they had constructed only a few feet from breeding nests of a Cactus Wren in the same chollas. The pyracantha bush at our front door, in 1961, contained an Inca Dove's nest and two Cactus Wren roosting nests. At least one of the latter was occupied.

House Sparrows lived in our neighborhood during the entire period of our study. They offered a good opportunity to observe the impact of an introduced species on resident desert birds. The food supply had to be shared; the sparrows found abundant insect food in the upper parts of the mesquite trees in the summer months. The wrens searched the lower part, gathering their food from the trunk and larger branches. Both species foraged on the ground.

House Sparrows avoided the cholla cacti; the adults could not be induced to land or perch on the spiny joints, or on roosting nests in these shrubs, even if we placed food there. A few of their inexperienced fledglings flew into chollas and became impaled on the spines and died. Obviously, we observed no competition for cholla nesting sites. The sparrows built their nests under the eaves of small buildings, or occasionally in catclaw bushes. An abandoned Cactus Wren's nest in a catclaw bush was appropriated by House Sparrows with no apparent difficulty. Another nest, in a large pyracantha shrub, was similarly occupied, but later the wrens took possession again. We seldom observed chasing.

At other times, more serious conflicts occurred. Our next door neighbor reported that in April 1942, a Cactus Wren entered the nest of a House Sparrow and removed at least one of the eggs, despite the spirited attack of the owners. The Cactus Wrens' young had been fledged 11 days earlier, and their presence in the vicinity may have contributed to the defensive raid. On 27 March 1954, we saw a male wren raid the House Sparrow nest in our backyard bird box. This bird box had been installed as a control to enable us to compare nesting dates. House Sparrows occupied it regularly year after year, both for roosting and for breeding purposes, without any violent interference. At this particular time, the female wren had been found dead in her nest, a short distance away, a few days before. In his efforts to obtain a new mate, the male sang from every point in the territory, moving about vigorously, apparently challenging the world. We found a dead House Sparrow nestling on the ground that morning. Soon afterward, we examined the nest box; it contained only one egg. That afternoon we saw the male wren fly to the nest box, and perch in the hole while he looked inside. Then he pulled out a feather. Half a dozen House Sparrows now gathered around, excitedly uttering their alarm notes. One flew at the wren and both tumbled to the ground, fighting furiously. The wren drove the sparrow aside and returned to the box. Soon it came out with the egg in its bill. It flew to the ground, dropping the egg; then it pecked it once. Amid the clamor of the sparrows, the wren returned to the nest, evidently to explore it further. When he left the box, he again fought with the sparrows.

In the course of the next half hour, the wren entered the box three more times. Once he sang from the top of the bird box, immediately engaging in a fight with a male House Sparrow. They fought in mid-air, 3 feet (.9 m) above the ground, the sparrow holding the upper position. Finally they dropped into a creosote bush, the sparrow 2 feet (.6 m) above the wren, chattering violently, but refusing to budge. The wren then left, apparently the loser in the brief battle. Later, however, he visited the box again to pull out a feather, which he dropped to the ground; then he chased a sparrow. This done, he flew northward and sang. In a few moments, the male sparrow returned to its perch on the box. On the following day, the wren again removed a feather from the nest. This time he carried it to his secondary nest. The loss of his mate, the complete break in the breeding pattern, and the frantic effort to secure his territory, may have been responsible for the extreme and unusual behavior.

In late summer, House Sparrows gathered in flocks and extended their feeding activities into adjacent areas. Large numbers roosted in ornamental bushes in the vicinity. In general, they filled a niche within the Cactus Wren territory which was not used by wrens.

House Finches were present in our vicinity at least as far back as 1934. They roosted usually in cholla cacti, sometimes beneath roosting nests of Cactus Wrens, sometimes in old open nests, or even near breeding nests. They were chased occasionally at roosting time if the wren happened to find the House Finch already in the cholla. They built their nests in chollas in March and April, in direct competition with the wrens. In fact, we gained the impression that they chose only the sites which the wrens permitted them to have. We recorded nesting attempts in most of the years from 1934 to 1946. Five of these were successful; in two of the years, two pairs succeeded in fledging young. While House Finches were seen frequently in the vicinity in the course of the past 12 years, we observed no further nesting attempts. Harassment usually consisted of chasing House Finches carrying nest materials. Once we saw a wren fly to a partly completed House Finch nest and tear loose some of the soft lining. Singing, interspersed with *scri* sounds, preceded this attack. At another time, we observed a wren sieze an egg in an abandoned House Finch nest. It flew with it to the ground and broke it.

Verdins (*Auriparus flaviceps*) did not breed in our study area. A single bird usually appeared in July or August and built a roosting nest in the large mesquite tree, or in the catclaw or pyracantha bushes. Verdins encountered no interference, either while nest building or while foraging in the territory, so far as we could observe. However, immature wrens in search of a roosting nest sometimes forced their way into one of these nests, distorting and damaging it so severely that the Verdin abandoned it.

On the Santa Rita Experimental Range, Cactus Wrens and Curve-billed Thrashers shared the habitat with breeding Gambel Quail (*Lophortyx gambelii*), Western Kingbirds, Ash-throated Flycatchers (*Myiarchus cinerascens*), Verdins, Black-tailed Gnatcatchers (*Polioptila melanura*), House Finches, Rufous-winged Sparrows (*Aimophila carpalis*), and probably Brown Towhees, all of which had little or no effect on the nesting of the wrens.

In Saguaro National Monument, along with the wrens and thrashers, 14 other permanent resident species of birds and five summer visitants nested within our research area or in its immediate vicinity (Table 19.1). All 19 lived more or less amicably with their neighbors. The Gilded Flicker (*Colaptes chrysoides*) and Gila Woodpecker, the Sparrow Hawk (*Falco sparverius*), Elf Owl (*Micrathene whitneyi*), Ash-throated Flycatcher (possibly Wied's Crested Flycatcher, *Myiarchus tyrannulus,* also), and Purple Martin (*Progne subis*), avoided nesting conflicts with the wrens and thrashers (but not with each other) by digging or selecting holes in saguaros. Gambel Quail, normally ground nesters, infrequently and inexplicably laid eggs in the nests of the Curve-billed Thrasher and Roadrunner, but to no avail, for the nests were deserted. Both Dawson (1923: 1588) and Brandt (1951: 137) have reported similar layings in thrasher nests. The sparrows nested in low shrubs; House Finches chose cholla cacti; Phainopeplas, palo verde trees; and Brown Towhees and Mockingbirds, both chollas and palo verde. The two species of doves and the Roadrunner were as versatile as the wrens in their choice of nest locations. Verdins, nesting chiefly in palo verde, but sometimes in cholla cacti, apparently suffered considerable losses. Their abandoned, disintegrating roosting nests were a common sight. Black-tailed Gnatcatchers

Table 19.1. Other species of birds breeding in the Saguaro Monument area.

Permanent resident	Est. no. of pairs
Sparrow Hawk **(Falco sparverius)**	1
Gambel Quail **(Lophortyx gambelii)**	4
Mourning Dove **(Zenaidura macroura)**	6
Roadrunner **(Geococcyx californianus)**	1
Gilded Flicker **(Colaptes chrysoides)**	3
Gila Woodpecker **(Centurus uropygialis)**	4
Verdin **(Auriparus flaviceps)**	3
Mockingbird **(Mimus polyglottos)**	2
Black-tailed Gnatcatcher **(Polioptila melanura)**	1
Phainopepla **(Phainopepla nitens)**	4
House Finch **(Carpodacus mexicanus)**	3
Brown Towhee **(Pipilo fuscus)**	2
Rufous-winged Sparrow **(Aimophila carpalis)**	2
Black-throated Sparrow **(Amphispiza bilineata)**	2
Summer resident	
White-winged Dove **(Zenaida asiatica)**	8
Elf Owl **(Micrathene whitneyi)**	1
Ash-throated Flycatcher **(Myiarchus cinerascens)**	2
Purple Martin **(Progne subis)**	4
Bendire Thrasher **(Toxostoma bendirei)**	1

nesting in palo verde trees dove repeatedly at approaching Cactus Wrens, apparently regarding them as serious enemies. House Finches and Mourning Doves succeeded in fledging young from nests as close as 2 feet (.6 m) from thrasher breeding nests. Each spring a Bendire Thrasher (*Toxostoma bendirei*) sang in the vicinity, but it was not until 1968 that a pair appeared with a fledgling from its extralimital nest.

Winter groups of Cactus Wrens, sometimes of as many as seven individuals, moved slowly about, feeding on the ground as they trespassed on neighboring territories, or joined their occupants briefly. Usually with them were one or two Curve-billed Thrashers, Brown Towhees, Black-throated Sparrows, Gila Woodpeckers, and Gilded Flickers. Small numbers of winter visitants, such as Brewer Sparrows and White-crowned Sparrows accompanied the group, sharing the food supply.

Predation

The most dangerous predator in the Kleindale Road vicinity was the house cat. We wondered at times how the birds could hold on at all, for almost every home had a cat which prowled at will. Feathers, chiefly from Cactus Wrens, were a common sight along the fences. Small boys with BB guns took their toll also. Snakes rarely visited our lot, and no losses could be attributed to them. Occasional Sparrow Hawks did not attack the wrens; Cooper Hawks (*Accipiter cooperii*) and Sharp-shinned Hawks (*A. striatus*) were absent.

Cats, Roadrunners, and snakes, even dead snakes, excited the wrens as soon as they were discovered in the territory. Uttering the *buzz* danger note persistently, the wrens gathered around the intruder and followed it as it moved away. House Sparrows joined the group and chimed in with their warning calls. Evidently enemy discrimination was accurate, for our small brown dog was completely ignored as he roamed the yard.

Loggerhead Shrikes (*Lanius ludovicianus*) nested in the Rillito Creek bottomlands a quarter of a mile (0.4 km) away. Now and then they appeared in our neighborhood. So far as we could determine, they never molested the wrens. No warning *tek* or *buzz* note greeted them to indicate to us that they were considered to be enemies, and no mobbing occurred. Yet, they were never permitted to remain very long. The moment a shrike arrived on a bush, post, or electric wire, a wren flew toward it, usually stopping about 10 feet (3 m) away. They eyed each other; then the wren moved closer with a fidgety, threatening motion. At 3 feet or so (0.9 m), the shrike seemed undecided. Finally it retreated a short distance. Then the wren repeated the maneuver. Sometimes it had to fly directly at the shrike to emphasize its threat. The shrike did not always give way. When it fought back, the wren retreated. Back and forth they fenced, the wren never giving up, the shrike retreating a little more each time.

One evening just before roosting time, such a seesaw battle lasted for 15 minutes, in and out of the creosote bushes and around them on the ground. The shrike climbed and twisted its way through the branches; the wren darted faster in a circle until it was hounding the other again. As it grew darker, the shrike departed without having made contact with the wren. The latter then retired into its nest.

We rarely sighted accipiter hawks on the Santa Rita Experimental Range. A Red-tailed Hawk (*Buteo jamaicensis*) nested in the vicinity. A pair of Roadrunners nested in territory 4 in 1953; we saw them occasionally in other years. In view of the wrens' ready recognition of this species as an enemy, it seems improbable that it succeeded in capturing many of the immature wrens. A Great Horned Owl (*Bubo virginianus*) roosted in the trees along the main wash; Screech Owls (*Otus asio*) were present — we found one in the cattle tank, apparently drowned. Shrikes visited the area, but they did not nest there. None of these carnivorous species, or indeed all of them combined, would appear capable of causing any great reduction in the Cactus Wren population during the winter. Perhaps in 1953 the carrying capacity of this part of the range had been exceeded, and the following years should be regarded as the normal ones. The required reduction in numbers may have been accomplished by disease.

As usual, the evidence of predation is largely circumstantial. In the Saguaro National Monument snakes were seldom encountered. Racers (*Masticophis*) were observed only 11 times, Gopher Snakes (*Pituophis*), three times, and Rattlesnakes (*Crotalus*), four times. We believe these snakes were responsible for most of the losses of eggs and nestlings in cholla cacti. Their secretive nature, and abundant hiding places under grass and shrubs, made detection difficult. Racers and Gopher Snakes climbed cholla cacti without apparent discomfort. Once a Racer lay coiled in an old, flattened cup of a wren's nest. Robbed nests showed no damage indicating forced entry. Most nests in saguaro crotches could not be reached by snakes; a few were accessible from the branches of an adjacent palo verde tree. The heavier, slow-moving Rattlesnakes seem more adapted to searching rodent burrows for prey.

Two other hazards, both speculative, faced nestlings in saguaro nests. One came from the Gila Woodpecker, of which at least four pairs nested each year in holes in the giant saguaros in the research area. Dawson (1923: 1036) reported that the Gila Woodpecker conducts a "systematic search for birds' eggs, especially those of the Lucy Warbler, Yellow Warbler, and Arizona Least Vireo." None of these species nested in the Monument area, for it contained no riparian habitat. The Cactus Wrens did not utter alarm notes in the vicinity of these woodpeckers. Nevertheless, this bird is suspect, because it would be a simple matter for it to probe into a Cactus Wren's nest to devour the eggs or newly-hatched nestlings. On the negative side, if the Gila Woodpecker discovered that Cactus Wrens' nests contained food, one could reasonably suppose that many would be entered and searched. There is no evidence that it does this; Bent (1939: 254) adds nothing more. However, a light *peep* or a low rustle from a cavity must be attractive. Lack (1966: 34) reported that "Great Spotted Woodpeckers, *Dryoscopus major,* occasionally bored through the wood of the nesting boxes and ate the nestling tits (*Parus major*)."

The Great Horned Owl and probably also the Screech Owl constituted the second hazard. Horned Owls roosted singly in the daytime in shady crotches of the saguaros and in the denser portions of palo verde trees. They were flushed many times in the fall, winter, and spring. A pair nested in three of the years in a tall saguaro a mile (1.6 km) south of the research area, and in the other 3 years 0.5 mile (.8 km) to the east. One can imagine what would happen if an owl, flying in to a perch on a saguaro branch, settled down upon a wren's nest filled almost to overflowing with nestlings ready to fledge. Even in darkness there would be an immediate, frantic attempt to escape, but capture would be certain. An adult wren sleeping in a roosting nest would meet the same fate. Some breeding nests lay with the entrance facing upon the horizontal part of the branch, affording an easy perch for intruders.

Each year one or two pairs of Roadrunners nested in the area. They were probably incapable of capturing any important number of adult Cactus Wrens, or even the smaller fringillids. Small birds, upon detecting a Roadrunner, take to the shrubs and trees, uttering a chorus of alarm notes. In such a situation, the Roadrunner is unable to use its speed to any advantage. Stray, inexperienced fledglings on open ground, or weak, injured or diseased birds could be easier prey. Red-tailed Hawks alternated with Horned Owls in the use of two nests in the saguaros, but they did not enter the research area during the course of the study. Cooper Hawks flew by occasionally; they did not nest in the vicinity. A Sparrow Hawk nested in a hole in a saguaro 0.2 mile (.32 km) south of the boundary. Its tendency to attack small birds has often been discounted. One which was surprised in flight dropped the headless body of a fledgling Mourning Dove.

Brown-headed Cowbirds (*Molothrus ater*), observed occasionally in the area, did not parasitize any of the Cactus Wren or Curve-billed Thrasher nests. Shrikes stopped by briefly and irregularly from August to January, perching on saguaros and palo verde trees. (On Kleindale Road, 15 January 1969, a Cactus Wren approached to within 3 feet (0.9 m) of a Shrike that was eating a House Sparrow wedged into a fork of a twig in a creosote bush. The Shrike suddenly dove at the wren, but the latter escaped easily, and was not pursued further.)

Bobcats (*Lynx rufus*) were rarely seen; Coyotes prowled the entire plot, revealing their presence by mournful barking in the early mornings. The two diurnal ground squirrels. Round-tailed and Antelope Ground Squirrels, both very common in the Monument, and frequently observed in chollas, did not molest the wrens. Little is known of possible depredations from nocturnal rodents. Once, in mid-morning, when a hand was inserted into an unfinished Cactus Wren's nest, a large Wood Rat (*Neotoma albigula*) popped out, and sought cover beneath the nest among the cholla twigs.

Losses in the Monument and the Kleindale Road locality were each approximately one-third of the clutches laid. Evidently the Cactus Wren's most dangerous period is not in the egg or as a nestling. It occurs after fledging. In the earlier study, we attributed the loss of fledglings to the ever-present house cat, especially during the day when the fledglings were more vulnerable. The diurnal predators in the Monument seemed unimportant. Failure of adult wrens to lead all of their fledglings to the roosting nest — a not so uncommon occurrence — would increase the exposure to nocturnal enemies. Inexperience in locating food may be the principal cause of juvenile mortality. Weak birds would soon succumb.

20. Discussion

Our research plot in the Santa Rita Experimental Range lay at the indefinite border where the desert scrub intruded into the grassland. There was no discernible change in the vegetation there in 1970. According to Humphrey (1953: 164), the desert grassland is not a true climax, but a subclimax maintained by fire. With fires now under control, the true climax of shrubs and cacti is gradually appearing. The Cactus Wrens' future seems assured here. Their population density from 1953 to 1956, excluding the extralimital territories 7 and 9, varied from 8.3 to 23.3 pairs per 100 acres (40.5 ha). The average is 13.3 pairs, but the high figure of 1953 is probably exceptional. The Curve-billed Thrashers' density varied from 5 to 8.3 pairs per 100 acres, and it was always lower than that of the wrens in the same year. The average was 6.3 pairs.

The Saguaro National Monument study area can be considered an optimum habitat for both Cactus Wrens and Curve-billed Thrashers. The Cactus Wren population density, calculated per 100 acres (40.5 ha), varied from 18 to 38 pairs, the Curve-billed Thrasher from 10 to 16 pairs, the averages being 26.7 and 13.3 pairs per year respectively. These population densities are far higher than any reported by other workers on these species in our southwestern desert scrub. They were higher, too, with the possible exception of the White-winged Dove, than that of any of the other 19 species which nested in the area. The number of territorial pairs of wrens and thrashers varied roughly in parallel in the course of the first 4 years, but in 1967, when the wrens rebounded from their low of 1966, the thrashers failed to gain. In 1968, both species held at the 1967 level of territorial pairs. The ratio of wrens to thrashers, averaging 2:1, reached 2.37:1 in 1965, the year of maximum population. Then followed a 52.6 percent drop in the wren population; the thrashers lost only 12.5 percent, and the ratio of wrens to thrashers decreased to 1.28:1; in 1967 it rose to 2.28:1. Although the densities of the two species were obviously in a continual, variable state, there is no proof that one species was crowding out the other. The 72 percent increase in the wren population, 11 to 19 pairs, from 1964 to 1965, accompanied a 33 percent increase, 6 to 8 pairs, in the thrasher population.

Table 20.1 compares the reproductive abilities of Cactus Wrens and Curve-billed Thrashers. Although the thrashers averaged more clutches per territory, their average eggs per clutch was lower. All hazards considered equal, the wrens' average of about 12 percent more eggs per pair each year must be an important advantage in maintaining their population ratio of 2:1 over the thrashers.

Systematists have long regarded the Troglodytidae and the Mimidae as closely related. Grinnell, in his study of two species of chickadees (1904: 376-377), concluded: "It is only by adaptations to different sorts of food or modes of food gathering, that more than one species can occupy the same locality. Two species of approximately the same food habits are not likely to remain long evenly balanced in numbers in the same region. One will crowd out the other . . ." Whether the wrens and thrashers can qualify for this close relationship is uncertain. They have "approximately" the same food habits, with the advantage on the side of the thrasher with its wider utilization of the habitat's food resources (except for upper level shrub and tree insects), and its ability to extract items from beneath the ground surface. This advantage alone would appear to be sufficient proof that these different population densities are not the result of competition for food.

Despite apparently excessive nesting losses in the Monument area, Curve-billed Thrashers maintained their population level with only minor fluctuations. Variations in spring temperatures and winter rainfall influenced the date of laying, but not the size of the territorial populations. Autumn and winter mortality apparently determined the latter. Hensley's (1954: 204) report of nesting success in the 1949 season in the Organ Pipe Cactus National Monument in southwestern Arizona shows that Cactus Wrens fledged a remarkable 83 percent of the eggs laid, and Curve-billed Thrashers 72 percent. The breeding pairs per 100 acres

Table 20.1. Comparative reproduction of Cactus Wrens and Curve-billed Thrashers, Saguaro National Monument, 1963 to 1968.

	Cactus Wren	Curve-billed Thrasher
Average territorial pairs	13.3	6.7
Average clutches per territory	2.0	2.2
Average eggs per clutch	3.3	2.7
Average eggs per pair per year	6.6	5.9
Total eggs in six years	514(99)	229
Percentage eggs hatched	82.5	54.6
Percentage hatch fledged	78.3	38.4
Percentage eggs fledged	64.6	20.9
Percentage successful clutches	68.8	24.4

(40.5 ha) calculated from the total of his three study areas were 8.1 Cactus Wrens and 4.5 Curve-billed Thrashers, resulting in an unbalanced ratio of 1.8:1. Whereas Hensley's study, including only one breeding season, may not be strictly comparable to a 6-year record, his figures do lend support to the supposition that the ratio of wrens to thrashers was independent of their nesting success. The high thrasher success in the Organ Pipe study occurred when the population disparity was nearly 2 to 1. The low ratio of 1.28:1 in 1966 in the Saguaro Monument was a result of the severe decline in the Cactus Wren breeding population that year.

A number of behavioral differences appear to favor the Cactus Wren. Its breeding nest effectively conceals the occupants from nocturnal, avian enemies, and, after the nestlings have fledged, it becomes their sleeping quarters for the first critical weeks. In winter, the roosting nest provides shelter from the cold, and to some extent from the rain. Occasionally, at the beginning of the nesting season, the wrens lay in roosting nests instead of constructing new nests. Secondary nests, built by the male while the female incubated, usually become the home of the next brood. The roosting nest must be of prime importance in establishing and maintaining faithfulness to a definite territory. It serves as a prominent, recognizable landmark, probably as individual as the configuration of the landscape.

Lack's (1966: 273) conclusion that ". . . the immediate factor determining the start of breeding is the female's obtaining enough food to form eggs" would appear applicable to Cactus Wrens and Curve-billed Thrashers. Early laying occurred in 4 of the 6 years when adequate rainfall, coupled with rising temperatures, produced an abundant growth of ephemeral plants and their associated insects. The years 1964 and 1967, however, were dryer (Table 8.4) and the wrens laid late; the correlation is less consistent in the thrashers. There can be no question of the advantage of being able to predict the food availability a month in advance in order to adjust the date of laying and clutch size to correspond to the most favorable conditions for the nestlings. No evidence whatever in this study confirms that either the wrens or the thrashers have such an ability. Food is a matter of chance, except that in normal years all went reasonably well.

Both species begin incubation before their clutches are complete, and asynchronous hatching is the rule. Lack (1966: 224), amplifying an earlier statement, says: "On present evidence I see no reason to change my view that the function of asynchronous hatching in the White Stork and other species, is to bring the effective brood-size rapidly down to that which the parents can adequately feed." Ricklefs (1965: 505-510), too, has accepted this view, and applies it to the Curve-billed Thrasher. It is difficult for one to escape the implication here that brood reduction is considered a deliberate act. The fact that in a "poor" year one or more of the younger nestlings dies is not proof that the parents deliberately favored the older nestings. Practically nothing is known of the manner in which food is presented to the nestlings, or the quantity delivered at each visit, or the number of young which receive food. It is improbable that any begging nestling is completely ignored. If a nestling fails to respond to the offer of food, it may be from weakness, crowding by its neighbors, congenital defects, or even from injury caused by its brooding parent, whose weight is 20 times that of the newly hatched bird. All eggs do not hatch; not all offspring are born healthy and hungry. Human experience will attest to that! Ricklefs' examples of brood reduction in June and July have to be taken into consideration of extreme environmental temperatures. The cumulative effect, of course, of any of the above anomalies is a diminished ability to beg for food, resulting in eventual starvation.

In a multi-brooded species such as the Cactus Wren in southern Arizona, whose egg-laying extends into June and sometimes July, it becomes imperative to begin incubation after the first egg is laid. In these months, a delay of from 2 to 4 days until the clutch is complete would be disastrous to the development of the embryos, because of the lethal mid-afternoon temperatures. Baldwin and Kendeigh (1932: 141) found "that in the case of eggs, young birds, and adults of the house wren, body temperatures between 114F (45.6C) and 116F (46.7C) were absolutely fatal." Willis (Auk 87: 827), reviewing Ricklefs' "An analysis of nesting mortality in birds" (Smiths. Contr. to Zool., 9: 1-48), lists a number of possible alternative causes of nestling mortality, and remarks that "Perhaps starvation is often the cause, but one must be cautious."

It is possible that in southern Arizona, at least, early incubation by the Curve-billed Thrasher could also have evolved in order to protect the eggs from the direct heat of the sun. Thrasher nests are seldom in complete shade. Afternoon temperatures, even in March, are often high, and direct sunlight on the eggs in the open nest could be injurious. In late nests it would be lethal. Since both sexes share the duties of incubation, the eggs are always shaded. The female Cactus Wren sleeps in her breeding nest, thus warming the first egg laid. Once begun, incubation logically continues, interrupted only by short absences to obtain food. The covered nest protects the eggs from the sun; assistance by the male is not required. To the thrasher, early incubation may increase slightly the risk of loss, for an occupied nest is probably more readily detected by predators.

By awakening earlier and retiring later than the wrens, thrashers increase their exposure to predation by owls in the twilight. Their food requirements must be greater because of their longer period of activity. When the spring ephemeral plants have run their course and dried, the wrens can forage in tree foliage for additional food, but the thrashers are restricted to the ground, or to picking small fruits.

The inexplicable destruction of Cactus Wren nests, reaching its maximum in autumn, may do some harm to juvenile wrens that must temporarily remain in exposed situations at night. On the whole, competition between the two species is negligible. Neither the Cactus Wren nor the Curve-billed Thrasher have any behavioral attributes which can be regarded as seriously inimical to the other.

When the Cactus Wrens established nine new territories in the population explosion of 1965, they filled in a number of little used areas and crowded into the boundaries also of the 11 territories of the previous year. Nesting success remained good, with no appreciable interference from disrupting conflicts. Thrasher territories, on the contrary, were larger, and they changed very little. Numerous conflicts occurred, indicating that their territories could not be compressed like those of the wrens to make room for additional pairs. Adults were long-lived; fledgling losses were apparently severe. Earlier (Anderson and Anderson, 1963: 36), we suggested that "the thrasher requires a larger territory, and its lower population density is a direct result of its own intraspecific aggressiveness." This view can now be repeated with more certainty. The Curve-billed Thrashers appear to be victims of their own intraspecific intolerance. This intolerance and their losses limited their population density to a low level in the Saguaro National Monument research area.

The destruction of the Cactus Wren's habitat proceeds at an alarming rate in the vicinity of the ever-expanding southwestern desert cities. Never, however, can one imagine that all of the desert will at some time be gone. It is too extensive, and it can doubtless always provide a reservoir of emigrant birds. If given a few ornamental cacti, the hardy, adaptable, aggressive Cactus Wrens will probably be able to maintain themselves within the residential sections of even the larger cities. In Tucson, the wrens can now be found nesting in olive, eucalyptus, and palm trees. Some years ago, Milam Cater reported to us that he was successful in attracting them to bird boxes at his home. If House Sparrows could be prevented from usurping these substitute nest sites, we feel that the Cactus Wrens can become established as a dooryard bird. They are still nesting on Kleindale Road (1970). This interesting species has only one disappointing attribute — its so-called song.

Summary

We studied Cactus Wrens in our spare time over a period of about 30 years in the vicinity of our home on Kleindale Road near Tucson, Arizona. From 1939 to 1964, we trapped and color-banded a total of 46 adults or full-size immature birds and 55 nestlings.

From 1953 to May, 1956, we conducted a population study of Cactus Wrens on a 60-acre (24 ha) plot on the Santa Rita Experimental Range, 35 miles (56 km) south of Tucson. Since only week ends were available for visits, we attempted no trapping or banding.

Our third study area, from 1962 to 1968, was in Saguaro National Monument, 14 miles (22.4 km) east of Tucson, on a plot of about 49 acres (19.6 ha). Here we were able to trap and band 81 adult Cactus Wrens and 29 Curve-billed Thrashers, 389 out of an estimated total of 538 Cactus Wren nestlings, and all 108 Curve-billed Thrasher nestlings.

The Tucson area has a long hot season beginning in April and ending in October. Maximum temperatures above 90F (32.2C) are the rule from May through September. Half of the rainfall occurs usually between 1 July and 15 September, in the form of showers and thunderstorms. A second period, from December to March, has more general prolonged rainstorms. Precipitation has varied from 5.34 inches (135.6 mm) in 1953 to 17.99 inches (456.9 mm) in 1964; the mean was 11.17 inches (283.7 mm).

The Kleindale Road habitat, at an elevation of 2,400 feet (731.5 m), consisted of a creosote bush association with an irregular sprinkling of cholla cacti and a few catclaws and mesquites. The Santa Rita Experimental Range plot, elevation 3,300 feet (1,005 m), was a cholla "meadow," with blue palo verde along the washes and scattered mesquite, desert hackberry, and catclaw in minor invasions of the cactus meadow. In the Saguaro National Monument area, at 2,840 feet (852 m), the saguaro and foothill palo verde dominated the landscape, with scattered mesquites, catclaw, white thorn, desert hackberry, and Mexican crucillo in thickets. Cholla cacti of sufficient height to serve as nest sites averaged 15 plants per acre (0.4 ha) quadrat.

Cactus Wrens remained in the Kleindale Road vicinity during the entire year. We found no evidence of migration, or of flocking behavior. A pair of wrens, with some birds of the year and a few outsiders, usually formed a loose group, and occupied an area of approximately 15 acres (6 ha) in the winter. In the Monument, we have some observations of more extensive, temporary wanderings away from established territories.

All of the wrens required a covered roosting nest in all months of the year. When no old nests were available, the wrens built new nests. Cholla cacti were commonly used for nest sites on Kleindale Road and on the Range. Male and female nests of a resident pair were seldom far apart; they were sometimes in the same cholla. Bailey, at a higher elevation near the Santa Rita Mountains, found nests in mistletoe in catclaw and mesquite. In the Monument, roosting nests were in cholla cacti, in mistletoe in palo verde trees, in crotches in saguaros, and a few were in old, weathered saguaro stumps.

The usual design of the nest was that of a pouch with an entrance at one end. This form was "standard," but the materials varied with what was obtainable in the locality.

Difficulty in choosing a site was sometimes evident by the beginnings of nests left unfinished. Construction of a roosting nest might begin at any time of the day, but once started, work began early each following morning. Nests were occupied from 1 to 6 days after construction began.

Nest entrances faced outward from the cholla, evidently for ease in entry, and quick escape if necessary. Prevailing winds were not a factor in determining entrance direction.

The song is a simple, oft-repeated series of harsh sounds. There is a danger note, a warning note, a possible location note, and a boundary dispute note. A growl is heard during the recognition display. Nestlings and fledglings have begging notes.

Since the sexes are identical in coloration and size, sex discrimination must be by means of distinctive behavior. Pair formation has not been observed, so we can only conjecture at this point from the behavior of already paired birds. Males can no doubt recognize other males by means of their songs. Females are probably attracted to males upon hearing the song. When they meet, there is a threatening display by the male, including spreading of wings and tail, accompanied by a growling sound. The female also displays, and then crouches. A female thus reveals her sex by cowering. These displays occur throughout the year, and suggest the assumption that the male cannot recognize his mate at the beginning of the season except by threatening her each time they meet. Later the display may become a ritual in maintaining the pair-bond. Females probably recognize other females as such only when they are in conflict over a male.

Territorial intolerance began at least as early as January. By February 15, the area was usually cleared of other wrens. Females were most active in driving out other females by chasing and fighting. The male apparently kept other males out. Singing by the male increased as the territory was secured. Singing stations were most numerous in the vicinity of the roosting nests.

Disputes in the Kleindale Road area were frequent in 1947, after the young were fledged in territory I, and they continued until the end of February, 1948. Singing stations were not on the "boundary," but some distance inside. Disputes occurred when one pair or the other detected what appeared to be an intrusion upon its territory. Both pairs then faced each other along the "line" and moved slowly in parallel lines. Short songs and scratchy territorial sounds were uttered by both males and females, as they proceeded for 15 to 20 feet, after which they turned and retraced their steps. Threatening postures, such as fluffing out feathers and spreading the tail, were noted. Sometimes short chases and brief fights took place back and forth, but the "line" still held, and the battle was over in a minute or two. Then both pairs retreated. Fledglings participated in the disputes, and often apparently precipitated the quarrels by straying across the "boundary." Displacement behavior was evident as one Cactus Wren ran about, picking up nest material at the end of the dispute. Occasionally displacement activity consisted of threatening other species, such as Gila Woodpeckers and Pyrrhuloxias which happened to be near at the time of the dispute.

The Cactus Wren's territory is used for mating, nesting, and as a feeding ground for the young, and it is also retained as a roosting area for the remainder of the year. The female defended this territory only when her mate was present. Although the maintenance of a territory provided freedom from interference in the nesting cycle, the duties of nesting seemed to leave little time for boundary disputes. Territorialism probably assisted in maintaining the pair-bond; it limited, too, the number of pairs in the tract.

The basic instinct of self-preservation appears to manifest itself in an attitude of dominance, which in turn is expressed by maintaining ownership of a feeding and sleeping area.

Breeding territories in the Kleindale Road area varied from one to five, including those bordering the study tract. Destruction of the habitat resulted in only one territory remaining within the 10-acre area from 1948 to 1958. Encroachment upon the territory by other Cactus Wrens resulted in defensive behavior, but a compromise was reached by giving up some of the original land to the newcomers. The "boundary line" was well defined in 1947, when the two breeding nests were only 180 feet apart. On the Santa Rita Experimental Range, we found 16 first brood nests in 1953, 5 in 1954, 5 in 1955, and 8 in 1956. The Curve-billed Thrasher population did not fluctuate to a similar extent. In the Saguaro National Monument, the Cactus Wrens averaged 26.7 pairs and the Curve-billed Thrashers 13.3 pairs per 100 acres (40 ha). Yearly territories were 9, 11, 20, 9, 16, and 16 for the Cactus Wrens; the Curve-billed Thrashers, more stable, established 5, 6, 8, 7, 7, and 7 territories in the 49 acres (19.6 ha).

The autumn roosting nests seldom remained intact until the next year's breeding nest was begun. Nest destruction and change of nests were frequent. Curve-billed Thrashers destroyed roosting nests but not breeding nests.

The female probably chose the breeding nest site for the first brood, and the male then assisted her in the nest construction. On the Range, all breeding nests were in cholla cacti. In the Monument, the Cactus Wrens placed about 60 percent of their 6-year total of 154 breeding nests in cholla cacti; most of the remainder were in saguaro crotches or dead stumps. About 60 percent of the second nestings of double clutches and 76 percent of the third nestings in triple clutches occurred in saguaros. Nest orientation was random. Air movement through a nest depends not upon the direction faced by the entrance, but upon the resistance of the nest to air flow. Almost all Curve-billed Thrasher breeding nests were in cholla cacti.

Average time from beginning construction to the laying of the first egg was about 14 days. After the failure of a nesting attempt, the next egg was laid in 6 to 7 days. Copulation occurred as early as 18 days before the eggs were laid; the female always invited it, and indicated her readiness by crouching, singing, and quivering her wings.

Variation in time of laying the first egg was great. The date for breeding readiness probably has some genetic basis for a given population; it is subject to modification by important environmental deviations from the normal. Most of the layings in the Kleindale Road area occurred after a rise in temperature, or at above normal temperatures. In Arizona, below 3,000 feet (900 m) elevation, the first egg is usually laid in March. Early nestings occurred after mild winters with rainfall adequate for new spring plant growth. Estimated dates for the first egg are given for 22 years in the Kleindale Road area. In about two-thirds of these years, the layings occurred when the average of the mean temperatures for the preceding 7 days was between 50F and 60F (10C and 15.6C). Delays in nesting are attributed to failure to find a mate in time, loss of mate, or nest destruction. Rainfall by itself was not a factor inducing laying.

In the Range area, egg laying occurred a month later than at Tucson. The elevation is 900 feet (270 m) higher than the Tucson locality. Its maximum temperatures were 3F (1.7C) lower in the daytime; and because of inversion, the night temperatures were 3F (1.7C) higher. With average temperatures identical in the two places, we assume that egg laying was delayed at the Range, possibly because of the slower onset of warm spring temperatures there.

In the Monument, variations in date of laying the first egg were frequent, and often considerable in both wrens and thrashers. The thrashers usually laid earlier than the wrens. Winter-spring rainfall with rising temperatures apparently induced laying; the increased food supply was no doubt an important factor. When rainfall was low, both species laid late.

Cactus Wrens did not desert their nests when the entrances were disturbed and widened to facilitate inspection of the contents.

The females roosted in their breeding nests sometimes as much as 7 to 11 nights before the first egg was laid. Eggs were laid at the rate of one a day, on consecutive days in early morning. The average number of eggs per clutch was 3.41. Although as many as six clutches were laid in 1 year, the maximum number of broods raised was three. In four of the years on Kleindale Road, three broods were raised; in nine of the years, two broods were raised; and in four of the years, one brood was raised. The Monument Cactus Wrens averaged 2.0 clutches of eggs per territory per year; the thrashers, 2.2. In the 78 territorial seasons, 19 wrens laid only one clutch; 42 laid two clutches; and 17 laid three clutches. Nine thrashers laid only one clutch; 20 laid two; 9 laid three; and one laid 4 (incomplete, deserted clutches are excluded). The average length of the Cactus Wren nesting season was 3.9 months; that of the thrashers, 4.1 months. Cactus Wrens averaged 3.3 eggs per clutch, thrashers 2.7. In both species, most nests contained three eggs. The average eggs per pair per year was 6.6 for the Cactus Wrens, and 5.9 for the Curve-billed Thrashers.

Incubation was performed entirely by the female. Partial incubation began the night after the first egg was laid; daytime incubation was irregular until the clutch was nearly complete. The period of incubation was found to be 16 days. At one nest on the seventh day of incubation, there were 28 attentive periods averaging 14.8 minutes, and 29 inattentive periods averaging 11.7 minutes each. The range of variation was 1 to 28 minutes and 2 to 26.5 minutes, respectively. On that day, 54.9 percent of the time was spent in incubation. The day before the first egg hatched, the female devoted 50.3 percent of the time to incubation. The female frequently carried lining material to the nest when she came to incubate. Courtship feeding by the male occurred three or four times a day. At high afternoon ambient temperatures, the female probably sat on the eggs to cool them.

Hatching of the eggs was spread over a period of 2 to 3 days. In no nest did all the eggs hatch on one day. About 30 percent of the time was devoted to brooding during the first 3 days after hatching. The time decreased rapidly thereafter.

During the time the female incubated her eggs, the male began construction of one or more secondary nests. Some were begun as early as the day following the laying of the first egg; most were begun 8 to 14 days later. Sometimes these nests were placed close to the breeding nest in the same cholla; others were from 14 to 240 feet (4.2 to 72 m) distant. The female did not assist in their construction until after the young were fledged. She then laid her next clutch of eggs in the secondary nest. Variations in this "normal" behavior were rather common. Occasionally the female appropriated her mate's roosting nest for her next clutch. After a clutch failure, she sometimes ignored the secondary nests, and, with the help of the male, constructed an entirely new nest for eggs. Rarely she laid her next clutch in the same nest that housed the previous brood. Frequently these secondary nests were built as replacements of destroyed roosting nests. Three of the nests were built entirely by females whose breeding nests were too crowded with nestlings to permit comfortable night brooding; they should probably not be classed as secondary nests. None of the secondary nests could be termed "extra" or "decoy." All the completed ones served some useful purpose. Those which did not become breeding nests were used by the male, the female, or the fledglings for roosts.

The feeding rate when the last hatched nestling was 5 days old was: the female averaged 7.3 visits per hour; the male, 4.3 visits per hour.

Nestlings are fed small fresh insects from the time they are hatched. Feeding by regurgitation was not observed. Fecal sacs were carried away from the nests when the nestlings were from 5 to 8 days old.

As fledging approached, singing by the male Cactus Wren increased. The song appeared to be directed to the nestlings, and apparently served as a signal for them to leave the nest. In addition, adults induced nestlings to fly from the nest by stopping a short distance away and withholding food.

The average time nestlings spent in the nest was 20.9 days; the range was 19 to 23 days.

The first egg of the second clutch was laid in from none to 13 days after the fledging of the first brood; the average for 7 years was 6.8 days. The interval between second and third broods averaged 6 days; the range was 1 to 11 days.

The time required from laying the first egg to fledging of the young averaged 38.4 days for 14 broods; the maximum was 41 days, the minimum, 36 days. Normally in the Tucson region, the breeding season probably ends with the fledging of the last young in the middle of August. Our local birds varied the length of the season from 3 to 7 months.

Nestlings from five Cactus Wren nests in the Kleindale Road area were removed daily for observation, weighing, and measuring.

At hatching, down in varying amounts is present on the capital, spinal, humeral, alar, femoral, crural, and ventral tracts. By the age of 5 days, the sheaths of the primaries, secondaries, and their coverts have pierced the skin; at 7 days of age, the sheaths in all the tracts are out.

In another day, they have begun breaking open at the ends. The eye-slit opens at 5 days, but widens so slowly that vision is probably not effective for several additional days. At the time of hatching, the upper mandible is shorter than the lower; by the fifth day they appear equal, and at 9 days of age the upper is longer.

Sharp *peep* notes have been heard on the second day, and frequently thereafter for several days, but by fledging time this note has changed to the normal begging *dzip*. Strength increases rapidly; at 4 days the nestlings can rest on their tarsi, while their wings are used for braces. Fear reactions, such as backing up when an object is suddenly placed before them, were not observed until 11 days old. By 14 days, escape attempts were the rule. At the age of 6 days, temperature regulation apparently becomes effective.

The lowest weight on the day of hatching was 2.6 grams. Gain in weight is fairly uniform at first. By the 12th day, the curve of mean weight becomes flatter, in the manner typical for many other species of birds. Just before fledging, another gain is recorded.

The growth of the wings was slow up to the fifth day, at which time the sheaths pierced through the skin. Then the rate increased with the increments to sheath length until the 13th day, after which the rate dropped again. The tail feathers maintained a steady rate of increase.

Young Cactus Wrens attained the average adult weight of 38.9 grams at about 38 days. The average loss of weight per hour in trapped adult wrens kept overnight was 0.25 gram.

Adult wrens always endeavored to lead their fledglings to a roosting nest in the evening. By singing frequently and flying to the nest, they induced their brood to follow them. At first, the fledglings usually returned to their former nest; later they sometimes shifted to other nests. Those fledglings which failed to arrive at their roosting nests before dark were often lost.

Fledglings began moving into separate nests at ages from 52 to 70 days. Apparently individual aggressiveness caused them to seek new quarters.

The adults shared the feeding duties for a short while. Then, when the female began incubating her next clutch of eggs, the male took care of the brood.

Exploratory pecking occurred as soon as the wrens were fledged. Self-feeding began at about the age of 35 days. The fledglings begged from any wren which approached. They became independent of their parents at the age of 50 days, but they begged up to 14 days later.

The *buzz* danger note was heard from day-old fledglings. It seemed to indicate annoyance more than danger. The alarm note, the staccato *tek,* came soon afterwards. The subsong sometimes began at the age of 30 days, but usually it did not appear until some days later. It was a very ragged imitation of the adult primary song. Location notes, the *rack,* and especially the *tirrup* variation, came after the subsong began. In 3 or 4 months the subsong, through constant practice, became adult in quality.

Dust bathing is apparently innate. It occurs at the early age of 26 days, and is performed each evening before going to roost. Bathing in pools of water was infrequent.

The adults broke up the family bond by gradually ignoring the begging fledglings. Pecking and chasing were not common. Conflicts at roosting nests were more frequent, but, as a rule, juvenile wrens were not driven out of the territory during the breeding season.

Nest materials were picked up by fledglings at the early age of 32 days. The average age at which they carried lining materials to old nests was 62.8 days. New nests were started at the average age of 116.5 days.

We observed the adult-type recognition displays by juvenile wrens only twice. They occurred in August, and were probably abnormal.

Juvenile wrens sometimes permitted fledglings of the following brood to roost with them.

Two instances of juvenile wrens feeding fledglings of a later brood were observed.

On Kleindale Road, failure of clutches increased rapidly after the second clutch; all fifth and sixth clutches resulted in failure.

In Saguaro National Monument, successful clutches totaled 68.8 percent for Cactus Wrens, and 64.4 percent of the eggs fledged. Curve-billed Thrasher success was very low, only 24.4 percent, and 20.9 percent of the eggs fledged young. Nearly one-third of the 48 Cactus Wren nests that failed were robbed of their eggs; nearly two-thirds lost their nestlings. There were only two desertions. Thrashers deserted 22 nests; predators robbed 18; a cold rain killed 7 young nestlings. Cactus Wrens fledged an average of 4.3 young per pair, the thrashers 1.2. Most of the losses of eggs and nestlings are attributed to snakes. The destruction of Cactus Wren roosting nests by Curve-billed Thrashers apparently had no serious effect on the population. We found no evidence that thrashers practiced "brood reduction," as has been reported in the literature.

Cactus Wrens avoided the extreme high summer ground temperatures by seeking shade. In hot weather, they opened their bills slightly and raised their wings to permit freer air circulation. Water drinking by adult wrens occurred chiefly in fall and winter. Immature wrens drank in summer.

Daytime nest temperatures varied with the thickness of the nest roof. Thin-walled nests were hot, for the sun shone through; shaded, ventilated nests approached the standard air temperature.

Cactus Wrens retired at a light intensity about twenty times as great as the intensity when they left the nest in the morning.

The average age of seven males in the Kleindale Road area was

737 days; 16 females averaged 493 days. Some wrens were found dead in their roosting nests.

Five Cactus Wrens in the Saguaro National Monument area lived for at least 4 years; two thrashers were present for 6 years. Losses and replacements of adult Cactus Wrens were few in the breeding season. Thrasher losses are uncertain, for nobands could not be traced. The proportion of first-year Cactus Wrens in the population is estimated to have been at least 15.7 percent, probably higher. The lowest proportions of year-olds occurred in 1966 and 1968, when the average clutches per territory were the highest. None of the banded Curve-billed nestlings survived to breed in the area.

Whenever possible, a fledgling male Cactus Wren in the Monument remained in the immediate vicinity of the place in which it had been hatched. It dispersed only as far as it was forced by parental antagonism to go, usually into an adjacent area or open territory. If its parents were lost, it assumed control of the territory. Females moved farther to obtain mates, sometimes up to 1,400 feet (420 m), evidently being less attached to their original homes. Females losing mates usually left to find new ones in other territories. Two males without mates held territories, meanwhile building several nests. Territories of single females were not found. On Kleindale Road, mates were usually replaced rapidly after a loss. There seemed to be an adequate outside supply waiting to be called.

The Kleindale Road Curve-billed Thrashers nested in chollas, evidently competing for nest sites. Conflicts were numerous; each species, however, was successful in defending its own breeding nest and young. Thrashers frequently destroyed the roosting nests of wrens; they also destroyed breeding nests after they had been abandoned, or had become roosting nests. Most of the destruction took place from late summer to January. The wrens did not defend their roosting nests. Nests of the two species were spaced from 3 to 480 feet (1 to 144 m) apart. The Curve-billed Thrasher appeared to be a poor competitor. While it held its own in the territory, its productivity was lower than that of the wren. Generally, Cactus Wrens laid larger clutches and attempted more broods in a season. They succeeded in raising more young.

In the Monument, 40 (49.4 percent) of 81 winter-banded adult Cactus Wrens disappeared, but only 6 (20.9 percent) of the 29 winter-banded Curve-billed Thrashers vanished.

Cactus Wren territorial disputes occurred chiefly at the boundaries. Curve-billed Thrashers fought viciously in the vicinity of their nests, often chasing the defeated intruders for considerable distances. Wrens' territories were "compressible," thrashers' less so. Interspecific conflicts were not observed, although both species placed their breeding nests in cholla cacti.

Several behavioral differences appear to favor the Cactus Wren. The covered breeding nest protects the eggs from the sun; the occupants are concealed from enemies. The fledglings use the nest for sleeping quarters. Wrens and thrashers are in direct competition for insect food on the ground surface and the low vegetation within reach, but the thrashers also dig into the ground for buried arthropods. They supplement their diet with berries and cactus fruits. In the spring and summer the wrens forage in shrubs and trees for insects. The thrashers' feet are apparently not adapted for this type of gleaning.

We have no proof that Cactus Wrens are crowding out Curve-billed Thrashers. The wrens' average of about 12 percent more eggs each year, and their greater production of young, is but part of their successful biology. Their weaker intraspecific intolerance permits more territorial "compressibility" in years of surplus individuals. Their population density is greater, not only because their production is greater, but because the surviving surplus is allowed to crowd in to nest. The Curve-billed Thrashers' intraspecific intolerance is probably the cause of its comparatively lower population density. In the Monument, heavy nesting losses contributed to the disparate situation.

The 19 other species of birds nesting in the area had little effect upon the wrens and thrashers.

Appendix

I. Distribution of the Cactus Wren, *Campylorhynchus brunneicapillus*

The distribution of the seven recognized subspecies can be summarized as follows:

Campylorhynchus brunneicapillus anthonyi. Western Texas; southern New Mexico, in lower elevations; western and southern Arizona; southeastern California in Colorado and Mojave deserts to Inyo and Kern counties, and in arid parts of the coast from San Diego County to Ventura County; the southern tip of Nevada in Clark County near the Colorado and Virgin rivers; southwestern Utah in Washington County; Pacific Coast of Baja California south to Ensenada, and the Gulf Coast to about latitude 31°; northern Sonora and Chihuahua and northwestern Coahuila, Mexico.

C. b. brunneicapillus. Central and southern Sonora and northern coast of Sinaloa, Mexico.

C. b. bryanti. Baja California west coast in San Quintin district.

C. b. purus. Baja California, Mexico, east and west coasts from latitude 29° to 25°.

C. b. affinis. Baja California, Mexico, in Cape district south of latitude 25°.

C. b. seri. Tiburón Island, Sonora, Mexico.

C. b. guttatus. Mexican Central Plateau in Jalisco, Michoacán, southern Chihuahua, Durango, Zacatecas, Aguascalientes, Guanajuato, Querétaro, México, southern Coahuila, San Luis Potosi, Hidalgo, southern Nuevo León, Tamaulipas to southern and central Texas.

II. Chronology of the Nomenclature of the Cactus Wren, *Campylorhynchus brunneicapíllus*

1835. The Cactus Wren (*Campylorhynchus brunneicapillus*), the northern representative of an American genus, some of whose species extend as far south as Brazil, was first described in 1835 under the name of *Picolaptes brunneicapillus,* by a Frenchman, Frédéric de Lafresnaye, in Guérin's Parisian journal *Magasin de Zoologie.* Although Lafresnaye was intensely interested in American birds, he never traveled abroad, but was content to obtain the specimens for his collection by purchase from other individuals. In this case, his Cactus Wren came from Charles Brelay of Bordeaux, France, who apparently had acquired it from an officer of a ship returned from California. California at that time was an extensive, sparsely populated area under the rule of Mexico. The exact locality from which the specimen came will probably never be known. Lafresnaye suggested that it may have been collected in Peru, since the ship could also have stopped there. Much of the subsequent nomenclatural confusion arose from the failure to specify a proper type locality.

Superficially the specimen resembled the Woodhewers (Dendro-colaptidae). Lafresnaye confined his description entirely to plumage colors. He presented no details of structure. In the plate 47, the artist colored the back of the bird brown, the underparts very buffy brown, and the tail blackish brown with white bars and spots (de Lafresnaye, F. Le Grimpic à coiffe brune. *Picolaptes brunneicapillus.* Nob. (Pl. 47) in *Magasin de Zoologie.* Journal, Publié par F.-E. Guérin. A Paris Chez Lequien Fils. Libraire. Quai des Augustins N° 47. 1835. vol. 5. [pages 61-62, plate 47].)

1836. Gould described *Thyrothorus guttatus* from "Mexico." Again, the omission of a definite type locality contributed to the later confusion. (Gould, J. Exhibition of birds allied to the European Wren, with characters of new species. Proc. Zool. Soc. London. 1836, part IV, Oct. 11, 1836, pp. 87-90. Descr. on p. 89.)

1846. In a summary, Lafresnaye placed *Picolaptes brunneicapillus* and *Thryothorus guttatus* in the genus *Campylorhynchus,* along with nine other species. The text, in French, has numerous typographical errors. (de Lafresnaye, M. F. Mélanges Ornithologiques. Revue Zoologique par Société Cuvierienne. 1846. Tome IX. Anneé 1846, pp. 91-94. List on page 94.)

This arrangement was followed by Gray (The Genera of Birds. London 1844-49. 3 vols., in I, 1847: 159), Bonaparte (Conspectus generum Avium. 1850, vol. 1: 223), and Sclater (Proc. Acad. Natur. Sci., Phila. VIII, 1856: 263-265). Meanwhile Lawrence (Ann. Lyc. Natur. Hist., New York, 1852, vol. 5: 114) had described *Picolaptes brunneicapillus* from Texas, and Heermann (Jour. Acad. Natur. Sci., Phila. 1853, vol. II, second ser., Art. XXV, part III: 263) had reported it from Guaymas, on the Gulf of California.

1859. Xantus described *Campylorhynchus affinis* from Cape St. Lucas, Lower California (Proc. Acad. Natur. Sci., Phila., 1859: 298). In the same Proceedings, pages 301, 303-304, Baird considered *C. affinis* and *C. brunneicapillus* to be distinct species.

1864. In his "Review of American birds in the Museum of the Smithsonian Institution" Baird suggested that Lafresnaye's *C. brunneicapillus* and *C. guttatus* might be the same bird (Smiths. Misc. Coll., 1864, no. 181 [1]: 99-100).

1869. Dugés reported *Picolaptes brunneicapillus* from Guanajuato, Mexico (La Naturalese, 1869 [1]: 140) and Lawrence reported *C. guttatus* from Yucatan, Mexico (Ann. Lyc. Natur. Hist., New York, 1869, 9: 199).

1880. By this time the distribution of the Cactus Wren along the southern United States border had been fairly well worked out. Salvin and Godman began to question the applicability of Lafresnaye's name to this bird (Biologica Centrali-Americana. Aves, 1: 67).

1881. Sharpe named the bird of the southern United States *Campylorhynchus couesi.* He considered *C. brunneicapillus* to be *C. affinis* (Cat. Birds British Mus., 6: 196-197).

1888. Ridgway disagreed with Sharpe (1881) and reported that specimens of birds from the southwestern border of the United States "agreed exactly" with Lafresnaye's type (Proc. Boston Soc. Natur. Hist., 23: 383).

1893. Palmer reported that *Campylorhynchus* is preoccupied by *Campylirhynchus,* a genus of Coleoptera, and should be replaced by *Heleodytes* (Auk, 10: 86-87).

1894. The American Ornithologists' Union adopted Palmer's correction and replaced *Campylorhynchus* with *Heleodytes* (Auk, 11: 48).

Anthony described *Heleodytes brunneicapillus bryanti.* Range: southern California to northern Lower California. He shifted *Heleodytes affinis* to subspecific status, *H. b. affinis* (Auk, 11: 210-214).

1898. Nelson described *Heleodytes brunneicapillus obscurus* from the tableland of Mexico (Proc. Biol. Soc. Wash. 12: 58-59).

1902. In this major revision, Mearns restricted Lafresnaye's *brunneicapillus* to northwestern Mexico, *H. b. affinis* to the Cape St. Lucas area, *H. b. bryanti* to the upper peninsula and southern California, and *H. b. couesi* to Texas. He described a new subspecies *H. b. anthonyi* for the region between California and Texas. He recognized *H. b. obscurus* of the tableland of Mexico (Auk, 19: 141-145).

1904. Ridgway examined the type of *Picolaptes brunneicapillus* and decided it came from Guaymas or its vicinity. He could not separate *H. b. anthonyi* from *H. b. couesi*. He recognized *H. guttatus, H. b. brunneicapillus, H. b. bryanti, H. b. affinis, H. b. couesi,* and *H. b. obscurus* (Bull. U. S. Nat. Mus., 50 [3]: 516-524, 753-754).

Swarth, too, could not distinguish *anthonyi* from *couesi*. He also reported that *H. b. bryanti* did not range into California (Condor, 6: 17-19). Later, several authors agreed with Swarth that *H. b. bryanti* should be restricted to Lower California.

1930. Bangs agreed with Ridgway that *Picolaptes brunneicapillus* came from southern Sonora (Bull. Mus. Comp. Zool., 70 [4]: 313).

1930. van Rossem described *Heleodytes brunneicapillus purus* from Lower California (Trans. San Diego Soc. Natur. Hist., 6: 224-226).

1932. van Rossem described *Heleodytes brunneicapillus seri* from Tiburón Island, Sonora (Trans. San Diego Soc. Natur. Hist., 7: 120, 125, 138-139).

1934. Hellmayr regarded *H. b. obscurus* as a synonym of *H. b. guttatus,* now considered to be the form from the Mexican plateau. He described *H. b. yucatanicus* from Yucatan. He also recognized *H. b. couesi, H. b. bryanti, H. b. affinis, H. b. purus, H. b. brunneicapillus,* and *H. b. seri,* the latter mentioned in a footnote (Field Mus. Natur. Hist. Publ. 330, Zool. Ser. 13 [7]: 147-150).

1945. van Rossem restricted the type locality of *Campylorhynchus couesi* Sharpe to Ringgold Barracks, Texas. He regarded Sharpe's *couesi* to be merely a new name for *Picolaptes brunneicapillus* of Lawrence. *C. b. brunneicapillus* is in southern Sonora (Occ. Papers Mus. Zool. Louisiana State Univ., 21: 184-186).

1947. *Heleodytes* becomes *Campylorhynchus* again when the American Ornithologists' Union Committee on Classification and Nomenclature decided to follow the International Code (Auk, 64: 451).

1957. The new A. O. U. Check-list of North American Birds, following decision 84 of the Copenhagen International Zoological Congress in 1953, regarded words ending *rhynchus* as of neuter gender. Thus *C. brunneicapillus* became *C. brunneicapillum.*

In the same year, part II of the Distributional Check-List of the Birds of Mexico followed the old nomenclature endings. It recognized *C. b. couesi, C. b. bryanti, C. b. purus, C. b. affinis, C. b. seri, C. b.*

brunneicapillus, and *C. b. guttatus,* leaving the Yucatan Cactus Wren as a full species, *C. yucatanicus.*

1958. The International Commission corrected the erroneous ruling (no. 84) of the Copenhagen Congress, and again the gender of *Campylorhynchus* became masculine (Mayr, Auk, 75: 225). The second printing of the A. O. U. Check-List contained the corrections (Auk, 79: 493-494, 1962).

1960. The Check-List of the Birds of the World followed the earlier (1957) Check-List of Mexican Birds in listing seven races of the Cactus Wren, with *C. yucatanicus* a full species.

1964. Selander: "My treatment of the subspecies of *C. brunneicapillus* is similar to that suggested by Mearns (1902) except that I regard *C.* [*b.*] *couesi* Sharpe, a name currently applied to all birds from the southwestern United States and adjacent parts of México (A. O. U. Check-List Committee, 1957: 416; Miller et al., 1957: 151), as a synonym of [*C. b.*] *guttatus* Gould. Thus I extend the range of *C. b. guttatus* to southern and central Texas. Birds from western Texas west to California and in adjacent parts of México represent *C. b. anthonyi.*" (Univ. Calif. Publ. Zool. 74: 238). He recognized seven subspecies: *C. b. brunneicapillus, C. b. bryanti, C. b. purus, C. b. affinis, C. b. guttatus, C. b. anthonyi, C. b. seri.*

Literature Cited

Alcorn, S. M., and C. May.
1962.   Attrition of a saguaro forest. Plant Disease Reporter, 46: 156-158.

Allen, R. W., and M. M. Nice.
1952.   A study of the breeding biology of the Purple Martin (*Progne subis*). Amer. Midl. Natur. 47: 606-665.

Ambrose, J. E., Jr.
1963.   The breeding ecology of *Toxostoma curvirostre* and *T. bendirei* in the vicinity of Tucson, Arizona. Univ. Ariz., M. S. thesis, 40 pp., unpublished.

Anderson, A. H.
1934.   Cactus Wrens and thrashers. Bird-Lore 36: 108-109.

Anderson, A. H., and A. Anderson.
1948.   Observations on the Inca Dove at Tucson, Arizona. Condor 50: 152-154.

1957.   Life History of the Cactus Wren. Part I: Winter and pre-nesting behavior. Condor 59: 274-296.

1959.   Life History of the Cactus Wren. Part II: The beginning of nesting. Condor 61: 186-205.

1960.   Life History of the Cactus Wren. Part III: The nesting cycle. Condor 62: 351-369.

1961.   Life History of the Cactus Wren. Part IV: Development of nestlings. Condor 63: 87-94.

1962.   Life History of the Cactus Wren. Part V: Fledging to independence. Condor 64: 199-212.

1963.   Life History of the Cactus Wren. Part VI: Competition and Survival. Condor 65: 29-43.

1965.   The Cactus Wrens on the Santa Rita Experimental Range, Arizona. Condor 67: 344-351.

Antevs, A.
1947.   Cactus Wrens use "extra" nest. Condor 49: 42.

[217]

Armstrong, E. A.
1955.   The Wren. Collins, St. James Place, London.

Bailey, F. M.
1922.   Cactus Wrens' nests in southern Arizona. Condor 24: 163-168.

Baldwin, S. P., and S. C. Kendeigh.
1932.   Physiology of the temperature of birds. Sci. Publ. Cleveland Mus. Natur. Hist. 3: 1-196.

Banks, R. C.
1959.   Development of nestling White-crowned Sparrows in central coastal California. Condor 61: 96-109.

Beal, F. E. L.
1907.   Birds of California in relation to the fruit industry. U. S. Dept. Agric. Biol. Surv. Bull. 30. 100 pp.

Bent, A. C.
1939.   Life histories of North American woodpeckers. U. S. Nat. Mus. Bull. 174.
1948.   Life histories of North American nuthatches, wrens, thrashers, and their allies. U. S. Nat. Mus. Bull. 195.

Brandt, H.
1951.   Arizona and its bird life. The Bird Research Foundation, Cleveland.

Brown, H.
1888.   On the nesting of Palmer's Thrasher. Auk 5: 116-118.

Dawson, W. L.
1923.   The birds of California. Booklovers' ed., 4 vols. South Moulton Co.

Edwards, H. A.
1919.   Losses suffered by breeding birds in southern California. Condor 21: 65-68.

Enemar, A.
1958.   Om ruvningens igångsättande hos koltrast (Turdus merula). Vår Fågelvärld 17: 81-103.

Green, C. R., and W. D. Sellers, eds.
1964.   Arizona climate. Univ. Ariz. Press.

Grinnell, J.
1904.   The origin and distribution of the Chestnut-backed Chickadee. Auk 21: 364-382.

Heath, H.
1920.   The nesting habits of the Alaska Wren. Condor 22: 49-55.

Heermann, A. L.
1853.   Notes on the birds of California, observed during a residence of three years in that country. Jour. Acad. Natur. Sci. Phila., ser. 2, 2: 259-272.

Heinroth, O.
1922.   Die Beziehungen zwischen Vogelgewicht, Eigewicht, Gelegegewicht und Brutdauer. Journ. für Ornith. 70: 172-285.

Hensley, M. M.
  1954.   Ecological relations of the breeding bird population of the desert biome of Arizona. Ecol. Monog. 24: 185-207.
  1959.   Notes on the nesting of selected species of birds of the Sonoran desert. Wilson Bull. 71: 86-92.

Howell, A. B.
  1916.   Some results of a winter's observations in Arizona. Condor 18: 209-214.

Huey, L. M.
  1942.   A vertebrate faunal survey of the Organ Pipe Cactus National Monument, Arizona. Trans. San Diego Soc. Natur. Hist. 9: 353-376.

Humphrey, R. R.
  1953.   The desert grassland, past and present. Journ. Range Manage. 6: 159-164.

Jaeger, E. C.
  1922.   Denizens of the desert. Houghton Mifflin Co.

Johnston, R. F.
  1961.   Population movements of birds. Condor 63: 386-389.

Kendeigh, S. C.
  1934.   The role of environment in the life of birds. Ecol. Monog. 4: 299-417.

King, J. R.
  1955.   Notes on the life history of Traill's Flycatcher (*Empidonax traillii*) in southeastern Washington. Auk 72: 148-173.

Lack, D.
  1966.   Population studies of birds. Oxford Univ. Press.

Lanyon, W. E.
  1960.   The ontogeny of vocalizations in birds. Animal Sounds and Comm., 1960, publ. 7: 321-347.

Laskey, A. R.
  1946.   Some Bewick Wren nesting data. Migrant 17: 39-43.

Lowe, C. H., ed.
  1964.   The vertebrates of Arizona. Univ. Ariz. Press, Tucson, Arizona.

Miller, A. H.
  1936.   Tribulations of Thorn-dwellers. Condor 38: 218-219.

Nice, M. M.
  1937.   Studies in the life history of the Song Sparrow. I. Trans. Linn. Soc. N. Y. IV: 247 pp.
  1941.   The role of territory in bird life. Amer. Midl. Nat. 26: 441-487.
  1943.   Studies in the life history of the Song Sparrow. II. Trans. Linn. Soc. N. Y. VI: 1-328.

Nice, M. M., and R. H. Thomas.
  1948.   A nesting of the Carolina Wren. Wilson Bull. 60: 139-158.

Rand, A. L.
  1941.   Development and enemy recognition of the Curve-billed Thrasher *Toxostoma curvirostre*. Bull. Amer. Mus. Natur. Hist. 78: 213-242.

Ricklefs, R. E.
    1965.   Brood reduction in the Curve-billed Thrasher. Condor 67: 505-510.
    1966.   Behavior of young Cactus Wrens and Curve-billed Thrashers. Wilson Bull. 78: 47-56.
    1969.   An analysis of nesting mortality in birds. Smiths. Contr. to Zool. 9: 1-48.
Ricklefs, R. E., and F. R. Hainsworth.
    1968a.   Temperature dependent behavior of the Cactus Wren. Ecol. 49: 227-233.
    1968b.   Temperature regulation in nestling Cactus Wrens: The development of homeothermy. Condor 70: 121-127.
    1969.   Temperature regulation in nestling Cactus Wrens: The nest environment. Condor 71: 32-37.
Saunders, A. A.
    1956.   Descriptions of newly-hatched passerine birds. Bird-Banding 27: 121-128.
Scott, W. E. D.
    1888.   On the avi-fauna of Pinal County, with remarks on some birds of Pima and Gila Counties, Arizona. Auk 5: 159-168.
Selander, R. K.
    1964.   Speciation in wrens of the genus *Campylorhynchus*. Univ. Calif. Publ. Zool. 74: 1-259.
Shreve, F.
    1951.   Vegetation of the Sonoran desert. Carnegie Inst. Wash. Publ. 591.
Skutch, A. F.
    1935.   Helpers at the nest. Auk 52: 257-273.
Swanberg, P. O.
    1950.   On the concept of "incubation" period. Vår Fågelvärld 9: 63-80.
Tompa, F. S.
    1962.   Territorial behavior: the main controlling factor of a local Song Sparrow population. Auk 79: 687-697.
    1963.   Behavioral response of Song Sparrows to different environmental conditions. Proc. XIII Intern. Ornith. Congr.: 729-739.
Weather Bureau, U. S.
    1966.   Local climatological data, with comparative data for Tucson, Arizona, 1965. U. S. Dept. Comm.
Wheelock, I. G.
    1904.   Birds of California. A. C. McClurg and Co., Chicago.
Willis, E. O.
    1970.   [Review of Ricklefs'] An analysis of nesting mortality in birds. [In] Smiths. Contr. Zool. 9 (1969): 1-48. Auk 87: 826-828.
Wolford, M. J.
    1969.   Vocal repertoire of the Cactus Wren (*Campylorhynchus brunneicapillus*), Univ. Ariz. M. S. thesis, 46 pp., unpublished.

Index

Acacia constricta, 14
Acacia greggii, 7
Accipiter cooperii, 192; *A. striatus,* 192
Aimophila carpalis, 191, 192
Alcorn, S. M., and C. May, 65, 217
Allen, R. W., and M. M. Nice, 95, 217
Ambrose, J. E. Jr., 187, 217
Amphispiza bilineata, 40
Amsinckia intermedia, 9, 150, 187
Anderson, A. H., 169, 217
Anderson, A. H., and A. Anderson, xiii, 169, 189, 217
Anisacanthus thurberi, 12
Antevs, A., 129, 217
Aristida sp., 13, 187
Armstrong, E. A., 95, 218
Association: creosote bush, 7, 9; *Cercidium-Opuntia,* 12; cholla meadow, 11; giant cactus park, 13; saguaro-palo verde-cactus, 13; wash, 12-13; desert grassland, 196
Astragalus nuttallianus, 9
Athel, 7
Auriparus flaviceps, 191, 192

Baccharis brachyphylla, 12; *B. sarothroides,* 7
Bailey, F. M., 20, 22, 23, 29, 30, 202, 218
Baldwin, S. P., and S. C. Kendeigh, 95, 199, 218
Banding, methods of, 4-6

Banks, R. C., 118, 218
Beal, F. E. L., 150, 218
Bent, A. C., 22, 31, 66, 78, 79, 83, 85, 218
Blackbird, European, 91
Bobcat, 195
Boerhaavia caribaea, 9; *B. spicata,* 9; *B. wrightii,* 9
Bouteloua aristidoides, 9; *B. barbata,* 9; *Bouteloua* sp., 13, 16
Bowlesia incana, 9
Brandt, H., 11, 13, 22, 31, 32, 85, 86, 191, 218
Breninger, G. F., 78, 79, 86
Broom, desert, 7
Brown, H., 78, 218
Bubo virginianus, 193
Buckwheat, 27
Bunting, Lark, 40, 188
Burroweed, 12, 16
Buteo jamaicensis, 193

Calamospiza melanocorys, 40
California, 66, 150
Calliandra eriophylla, 12
Campylorhynchus brunneicapillus (see Wren, Cactus); *C. zonatus,* 138
Canis latrans, 184
Cardinal, 188
Carnegiea gigantea, 14, 22
Carpodacus mexicanus, 29
Catclaw, 7, 12, 14, 20, 189

Cater, M., 99, 200
Cat, House, 192, 195
Cattle, disseminate cholla cacti, 10; supervised on Santa Rita Experimental Range, 11; prohibited in Saguaro National Monument, 17
Centurus uropygialis, 4
Cercidium floridum, 12; C. microphyllum, 14, 22
Celtis pallida, 12
Chalk, J., 77
Chickadee, 197
China-berry, 7
Cholla, cane, 7, 20, 127; jumping, 7, 14, 83, 127; dissemination by cattle, 10; pencil, 22; staghorn, 7, 14
Citellus harrisii, 10; C. tereticaudus, 10
Climate, 17, 152, 201
Colaptes chrysoides, 191, 192
Condalia lycioides, 12, 21; C. spathulata, 14
Cottonwood, 7, 18-19
Cowbird, Brown-headed, 195
Coyote, 184, 195
Creek, Rillito, 7, 9, 10, 19, 69, 77, 148, 188
Creek, Tanque Verde, 19
Creosote bush, 7, 13, 14, 77, 79
Crockett, R. M., 79
Crotalus, 193
Crucillo, Mexican, 14, 187
Cryptantha angustifolia, 9; C. barbigera, 9; Cryptantha sp., 13, 187
Cynodon dactylon, 28

Dawson, W. L., 191, 194, 218
Descurainia pinnata, 9; D. sophia, 9
Desert, Colorado, 169; Mohave, 169
Dille, F. M., 78
Dipodomys merriami, 169
Dove, Inca, 188; Mourning, 149, 188, 189, 192, 195; White-winged, 192, 196
Dryocopus major, 194

Edwards, H. A., 169, 218
Empidonax traillii, 118
Enemar, A., 91, 218
Ephedra trifurca, 7, 9
Eriogonum deflexum, 9, 23; E. trichopes, 9, 27

Erodium, 13, 187; E. cicutarium, 9; E. texanum, 9
Erwinia carnegieana, 65
Evax multicaulis, 9

Fairy duster, 12
Falco sparverius, 191, 192
Festuca octoflora, 9
Filaree, 9, 76
Finch, House, 29, 40, 151, 188, 190-191, 192
Flicker, Gilded, 191, 192
Flycatcher, Ash-throated, 191, 192; Traill, 118; Wied's Crested, 191
Fouquieria splendens, 12

Geococcyx californianus, 192
Gnatcatcher, Black-tailed, 191, 192
Grama, needle, 9, 13, 187; six-weeks, 9, 16
Grass, Bermuda, 28; cotton, 13; mesquite, 13
Gray-thorn, 12, 187
Green, C. R., and W. D. Sellers, eds., 82, 218
Grinnell, J., 197, 218
Ground Squirrel, Antelope, 10, 169, 195; Round-tailed, 10, 139, 169, 195
Guaymas, 31

Habitat, Kleindale Road, 7-11, 13, 201; Saguaro National Monument, 13-17, 201; Santa Rita Experimental Range, 11-13, 201; changes, Kleindale Road, 9-11
Hackberry, desert, 12, 14, 187
Haplopappus tenuisectus, 12
Harrison, H., 77
Hawk, Cooper, 192, 195; Red-tailed, 193, 195; Sharp-shinned, 192; Sparrow, 191, 192, 195
Heath, H., 95, 218
Heermann, A. L., 31, 218
Heinroth, O., 92, 218
Hensley, M. M., 78, 92, 109, 173, 174, 219
Heteropogon contortus, 13
Honeysuckle, 12
Howell, A. B., 18-19, 219
Huey, L. M., 173-174, 219
Humphrey, R. R., 196, 219

Ironwood, 79

Jaeger, E. C., 127, 219
Johnston, R. F., 160, 219
Joshua tree, 24

Kallstroemia grandiflora, 9;
    K. parviflora, 9
Kangaroo rat, Merriam, 169
Kendeigh, S. C., 88, 219
King, J. R., 118, 219
Kingbird, Western, 188, 191

Lack, D., 194, 198, 219
Lanius ludovicianus, 193
Lanyon, W. E., 133, 219
Lappula redowskii, 9
Larrea divaricata, 7
Laskey, A. R., 95, 219
Lepidium lasiocarpum, 9
Lesquerella gordoni, 9
Lophortyx gambeli, 191, 192
Lotus sp., 16
Lowe, C. H., 80, 219
Lycium berlandieri, 7
Lynx rufus, 195

Martin, Purple, 191, 192
Masticophis, 193
Melia azederach, 7
Melospiza melodia, 5
Mesquite, 7, 14, 21, 77, 79, 186, 187,
    189
Mexico, 66
Micrathene whitneyi, 191, 192
Miller, A. H., 127, 219
Mimidae, 197
Mimus polyglottos, 188, 192
Mistletoe, leafless, 14, 187, 188
Mockingbird, 188, 191, 192
Molothrus ater, 195
Mormon tea, 7, 9, 12
Mountains, Beaver Dam, 24; Rincon,
    14; Santa Catalina, 14, 15, 77, 78,
    148; Santa Rita, 23, 29
Muhlenbergia porteri, 13
Mustard, bladder-pod, 9, 76-77, 79
Myiarchus cinerascens, 191, 192;
    M. tyrannulus, 191

Neotoma, 16; *N. albigula,* 195
Nevada, 17
Nice, M. M., 5, 47, 48-49, 72, 74, 79,
    110, 147, 184, 219
Nice, M. M., and R. H. Thomas, 95,
    219

Ocotillo, 12, 14
Opuntia arbuscula, 22; *O. engelmannii,*
    12, 16; *O. fulgida,* 7, 12, 22; var.
    mammillata, 12, 14, 22;
    O. phaeacantha, 16; *O. spinosior,*
    7, 12, 22; *O. versicolor,* 7, 22
Organ Pipe Cactus National
    Monument, 23, 78, 109, 173
Otus asio, 193
Owl, Elf, 191, 192; Great Horned,
    193, 195; Screech, 193

Palo verde, blue, 12, 14; foothill, 14
Parus major, 194
Passer domesticus, 4
Pectocarya platycarpa, 9;
    P. recurvata, 9
Phacelia crenulata, 9
Phainopepla nitens, 188, 192
Phillips, A. R., 77, 108-109
Phoenix, 78, 79
Phoradendron californicum, 14, 22
Pipilo fuscus, 188, 192
Pituophis, 193
Plantago insularis, 9; *P. purshii,* 9;
    Plantago sp., 16
Polioptila melanura, 191, 192
Populus fremontii, 7
Progne subis, 191, 192
Prosopis juliflora, 7
Pyracantha, 20, 189
Pyrrhuloxia sinuata, 47, 203

Quail, Gambel, 191, 192

Racer, 193, 194
Rand, A. L., 132, 219
Rattlesnake, 193, 194
Research locations: Kleindale Road,
    4; Saguaro National Monument, 5;
    Santa Rita Experimental Range,
    4-5; length of study, 3
Richmondena cardinalis, 188
Ricklefs, R. E., 117, 198-199, 220
Ricklefs, R. E., and F. R. Hainsworth,
    29, 67, 69, 106, 116, 149, 220
River, San Pedro, 78
Roadrunner, 174, 181, 184, 191, 192,
    193, 194-195

Sahuarita, 11
Salix sp., 7
Salpinctes obsoletus, 188
Saunders, A. A., 112, 220
Scardafella inca, 188

Schismus barbatus, 9, 27, 77
Scott, W. E. D., 78, 220
Selander, R. K., 124, 220
Shreve, F., 9, 12, 220
Shrike, White-rumped, 193, 195
Sisymbrium irio, 9, 77
Skutch, A. F., 138, 220
Snake, Gopher, 193, 194
Snakes, 174
Sparrow, Black-throated, 40; Brewer, 40, 188, 192; House, 4, 20, 25, 39, 40, 131, 187, 188, 189-190, 193, 195, 200; Rufous-winged, 191-192; Song, 5, 72, 74, 79; White-crowned, 4, 40, 118, 188, 192
Spizella breweri, 40
Stork, White, 198
Streets, R. B., 19
Stylocline micropoides, 9
Swanberg, P. O., 91, 92, 220

Tamarix aphylla, 7
Tanglehead, 13
Telmatodytes palustris, 112
Texas, 17, 66
Thorn, desert, 7
Thrasher, Bendire, 192
Thrasher, Curve-billed
    Breeding nest location, 180; competition for nest sites, Organ Pipe Cactus National Monument, 173-174; open, exposed nest, 149; distance between nests of Cactus Wrens and Curve-billed Thrashers, 171-173
    Conditions influencing start of laying, 181-182; delayed laying, 79; comparative dates, first egg, Kleindale Road, 173; date first egg, Saguaro National Monument, 181; clutches per year, 182-183, 205; length of season, 183, 205; interference by Gambel Quail, 191
    Conflicts with Cactus Wrens, Kleindale Road, 25, 40, 168-169, 173; in Saguaro National Monument, 185-186
    Destruction of Cactus Wren roosting nests, 39, 169-171, 204; erratic, 170; breeding nests not destroyed, 63; Cactus Wrens did not defend roosting nests, 170
    Food habits determined innately, 132; competition for food,

Kleindale Road, 168; food preference and feeding behavior, 186-187; Ambrose indicates they are almost omnivorous, 187
    Nesting success, Kleindale Road, 173; in Saguaro National Monument, 183; in Santa Rita Experimental Range, 174; in Organ Pipe Cactus National Monument, 197-198; predation, 184; comparative losses, 184; nestling and fledgling losses, 185; male survival, 184; dispersal, 185; comparative reproduction, 173; asynchronous hatching, 198
    Population density, on Kleindale Road, 168; in Saguaro National Monument, 196-197, 204; in Santa Rita Experimental Range, 174, 196, 203-204; loss of winter-banded thrashers, 175; unbalanced sex ratio, 180; polygyny, 180
    Territories, on Kleindale Road, 168; in Saguaro National Monument, 174-179, 204; conflicts and combat, 175, 180; territories larger because of intraspecific intolerance, 200
Three-awn, 13, 187
Tit (Great), 194
Tompa, F. S., 5, 220
Towhee, Brown, 188, 191, 192
Toxostoma bendirei, 192
Toxostoma curvirostre (see Thrasher, Curve-billed)
Tribulus terrestris, 9
Trichachne californica, 13
Tridens pulchellus, 27
Troglodytes aedon, 88; *T. troglodytes,* 95
Troglodytidae, 1, 197
Turdus merula, 91
Tyrannus verticalis, 188

Utah, 17

Verdin, 191, 192
Vireo, Arizona Least, 194

Warbler, Lucy, 194; Yellow, 194
Weather Bureau, U. S., 17, 152, 220
Wheeler, W. M., 48-49
Wheelock, I. G., 104, 220
White-thorn, 14
Willis, E. O., 199, 220

Willow, 7
Wolford, M. J., 33, 220
Woodpecker, Gila, 4, 40, 191, 192, 194, 203; Great Spotted, 194
Woodrat, 16, 169, 195
Woods, R. S., 22, 31, 66
Wren, Cactus
Arizona State bird, 1
Awakening, song, 41; time, 151, 208
Behavior, dust-bathing, 134-135, 208; water-bathing, 135; irrelevant, 140; winter, 18-19; favorable differences, 198-199, 209-210
Breeding cycle start, food a favorable factor, 198; copulation, 73-74, 204; ovum growth, 74
Chronology of the nomenclature, 212-215
Courtship feeding, 96, 103, 205
Description, 1
Distribution of the seven races, 211
Eggs, description, 83; rate laid, 84. Kleindale Road: clutch size, 85; total laid, 85; clutch size, relative number, 86. Saguaro National Monument: clutch sizes, 89, 205; clutches per territory, 87, 205; eggs per female, 89-90; composition of 1st, 2nd, and 3rd clutches, 90; percentage single, double, and triple clutches, 87; comparative production, 90
Fledging procedure, 106-107; rate of, 126
Fledgling roosting behavior, 127-129, 207; change of roosting nest, 129-130, 207; length of group roosting, 130; aggressive behavior, 130; exploratory pecking, 131, 207; parental duties, 131, 207; self-feeding, 131, 207; independence, 132, 207; development of vocalizations, 133-134, 207-208; boundary disputes, 134, 136; family bond break-up, 135-136, 208; juvenile display, 138; juvenile helpers, 138-140, 208
Food, animal, 150; vegetable, 150; water requirements, 149-150, 208; foraging in cottonwoods and date palms, 18-19; feeding manner, 186; on saguaro flowers, fruit, 187; assists in cross pollination, 187
Incubation, begins, 91, 205; by female, 91, 205; period, 92, 205; attentiveness, 92-95, 205; temperature dependence, 95, 205; in hot weather, 95, 205; nest materials carried to nest, 96, 205; time of hatching, 96, 205
Laying, climatic dependence, 74-78, 80-82; date first egg, Kleindale Road, 74-75, 77; in Phoenix, 79; in Saguaro National Monument, 80-82; in Santa Rita Experimental Range, 79; date second clutch, Saguaro National Monument, 88-89
Molt, 124
Nest, breeding, materials used, 66; construction time, 72, 204; constructed by both sexes, 71-72, 204; location choice by female, 69-71, 204; changes in location, 69-70; orientation of entrance, 66-69, 204; Kleindale Road nests, 60-63, 204; Saguaro National Monument nests, 63-66, 204; Santa Rita Experimental Range nests, 63, 204; nest first occupied by female, 84; nests in ornamental trees, 200; bird boxes, 200
Nest, roosting, always required, 19, 202; description, 22-24, 202; weight of, 28; site selection, 24-25, 202; entrance orientation, 29-30, 202; location on Kleindale Road, 20-21, 202; in Saguaro National Monument, 22, 202; in Santa Rita Experimental Range, 20-21, 202; construction, 26-29, 202; attentiveness in construction, 29, 202; date of occupancy, 26-27, 202; nest changes in suburban areas, 24, 28; construction by 1st year Cactus Wrens, 136-138, 208
Nest, secondary, use of, 101, 206; location, 99-100, 206; start of construction, 97, 205-206; constructed by male, 97-98, 205; by female, 99, 206; construction time required, 99-100; number of nests, Kleindale Road, 101; nests in Saguaro National Monument and Santa Rita Experimental Range, 101
Nesting season, length, Kleindale Road, 108-109; in Saguaro National Monument, 109-110;

226   INDEX

Wren, Cactus, Nesting season, cont.
other areas, 109; cause of
termination, 108-110
Nestlings, growth, 111-115, 206-207;
measurements, 123-124; weights,
118-123, 207; behavior, 115-118,
207; brooding time, 103, 205;
feeding routine, 102-106, 206;
feeding not by regurgitation, 104,
206; fecal sac removal, 105-106;
activity, 106; singing by male, 105;
nestling time in nest, 107-108;
interval between fledging and
next egg, 108; asynchronous
hatching, 198
Pair formation, 34, 202; male meets
male, 34; female meets male,
34-35; male meets female, 35,
202; female meets female, 35;
male with excess females, 36-38
Population density, Saguaro National
Monument, 192, 196-197, 204;
Santa Rita Experimental Range,
196, 203-204; winter population,
Kleindale Road, 18-19
Predation by House Cat, 192;
snakes, 193-194; Gila Woodpecker,
194; Horned and Screech Owls,
194; Roadrunner, 194-195;
wood-rats and ground squirrels
reported by Edwards, 169
Relationships with other birds on
Kleindale Road, 188-191; in
Saguaro National Monument,
191-192; in Santa Rita
Experimental Range, 191
Retirement, 150-151, 208
Sex determination, 6, 36, 202
Study methods, 3-6, 201
Success, hatching, Kleindale Road,
141-143; fledging, 142-143, 208;
clutch failures, 141, 143; nesting
success, Kleindale Road, 144, 208;
Saguaro National Monument,
144-145, 208; Organ Pipe Cactus
National Monument, 197-198;
percentage success 1st, 2nd, and
3rd clutches, Saguaro National
Monument, 145; annual
successful clutches, 145; success
of single, double, and triple
clutches, 146; young fledged per
pair, 147, 208; breeding nest
losses, 145; comparative
reproduction, 197, 208-209

Survival, 157; age attained, 152-153,
208-209; immature survival,
Kleindale Road, 152-153; dead in
nest, 152; breeding season losses,
replacements, 158; roosting nest
counts, 156-157; percent banded,
Saguaro National Monument,
154, 156; fledglings banded, 154;
percent 1st year birds in
population, 158; percent 1st year
males in population, 158; sex
ratio, 167; male dispersal,
160-162, 166, 209; female
dispersal, 162, 164, 166, 209;
nestling male, 164-165; nestling
female, 165; replacements,
159-160; loss winter-banded birds,
156; mortality from high
temperature, 199; starvation, 199
Territory, beginning of assertion,
38-39, 203; boundary disputes, 40,
43-46, 203; fledglings precipitate
disputes, 46; defense, 42-43, 203;
displacement or irrelevant
behavior, 46-47, 203; function of
territory, 47-49, 203; role of male,
38-39; role of female, 38-40, 45,
49, 203; female unable to hold
territory, 167; singing stations,
41-42, 203; size of territory, 51-54,
203-204; stability of, 54-55;
tolerance of other species, 40,
203; number of territories,
Kleindale Road, 50-52, 203; in
Santa Rita Experimental Range,
52-54, 204; in Saguaro National
Monument, 54-59, 204
Vocalizations, adult, 31-33, 41,
70-71, 202; display-growl, 35, 105,
202; nestling, 33, 202; fledgling,
33, 202
Weather, effect of precipitation,
151-152; of temperature, 148-149;
of windstorms, 151
Weight, adult, 125-126, 207;
fledgling, 125, 207
Wren, European, 95; House, 88, 112;
Long-billed Marsh, 112; Rock, 188

Yucca brevifolia, 24
Yuma, 78

Zenaida asiatica, 192
Zenaidura macroura, 149
Zizyphus, 21
Zonotrichia leucophrys, 4

5750